Mathematical Models in Finance

Mathematical Models in Finance

Edited by

S.D. HOWISON, F.P. KELLY

and

P. WILMOTT

Published by Chapman & Hall for The Royal Society

CHAPMAN & HALL

London · Glasgow · Weinheim · New York · Tokyo · Melbourne · Madras

Published by Chapman & Hall, 2–6 Boundary Row, London SE1 8HN, UK

Chapman & Hall, 2–6 Boundary Row, London SE1 8HN, UK

Blackie Academic & Professional, Wester Cleddens Road,
Bishopbriggs, Glasgow G64 2NZ, UK

Chapman & Hall GmbH, Pappelallee 3, 69469 Weinheim, Germany

Chapman & Hall USA, One Penn Plaza, 41st Floor, New York
NY10119, USA

Chapman & Hall Japan, ITP-Japan, Kyowa Building, 3F,
2-2-1 Hirakawacho, Chiyoda-ku, Tokyo 102, Japan

Chapman & Hall Australia, Thomas Nelson Australia,
102 Dodds Street, South Melbourne, Victoria 3205, Australia

Chapman & Hall India, R. Seshadri, 32 Second Main Road, CIT East,
Madras 600 035, India

First edition 1995

First published in 1994 by The Royal Society in Phil. Trans. R. Soc. Land.

Printed in Great Britain by Hartnolls Ltd, Bodmin, Cornwall

ISBN 0 412 63070 2

A catalogue record for this book is available from the British Library

Printed on permanent acid-free text paper, manufactured in accordance with ANSI/NISO Z39.48-1992 and ANSI/NISO Z39.48-1984 (Permanence of Paper).

Contents

Contributors

A. Bensoussan Domaine de Voluceau, Rocquencourt, BP 105, 78153, Le Chesnay Cedex, France

R.H. Brown London Business School, Sussex Place, Regents Park, London NW1 4SA, UK

J.M.C. Clark Department of Electrical and Electronic Engineering, Imperial College of Science, Technology and Medicine, London SW7 2BT, UK

M. Crouhy HEC School of Management, 78351 Jouy en Josas, France

M.H.A. Davis Department of Electrical and Electronic Engineering, Imperial College of Science, Technology and Medicine, London SW7 2BT, UK

J.N. Dewynne Department of Mathematics, University of Southampton, Highfield, Southampton SO9 5NH, UK

D. Duffie The Graduate School of Business, Stanford University, Stanford, California 94305, USA

H. Föllmer Institut für Angewandte Mathematik, Universität Bonn, D-53115 Bonn, Wegelerstrasse 6, Germany

D. Galai The Hebrew University, Mount Scopus, Jerusalem 91905, Israel

D. Harding Adam, Harding and Lueck Ltd, Willow House, Willow Place, London SW1P 1JH, UK

C.R. Hensel Research Department, Frank Russell Company, 909A Street, Tacoma, Washington 98402–1616, USA

S.D. Hodges Financial Options Research Centre, Warwick Business School, University of Warwick, Coventry CV4 7AL, UK

S.D. Howison Oxford University Mathematical Institute, 24–29 St Giles, Oxford OX1 3LB, UK

R. Kan The Graduate School of Business, Stanford University, Stanford, California 94305, USA

H.M. Markowitz 1010 Turquoise Street, Suite 245, San Diego, California 92109, USA

R.C. Merton Graduate School of Business, Harvard University, Boston, MA 02163, USA

S.R. Pliska Department of Finance, University of Illinois at Chicago, Chicago, Illinois 60607–7124, USA

S.M. Schaefer London Business School, Sussex Place, Regents Park, London NW1 4SA, UK

M.J.P. Selby Gonville and Caius College and the Faculty of Economics and Politics, University of Cambridge, Cambridge CB3 9DD, UK

A.E. Whalley Oxford University Mathematical Institute, 24–29 St Giles, Oxford OX1 3LB, UK

P. Wilmott Oxford University Mathematical Institute, 24–29 St Giles, Oxford OX1 3LB, UK and Department of Mathematics, Imperial College of Science, Technology and Medicine, London SW7 2AZ, UK

W.T. Ziemba Faculty of Commerce, University of British Columbia, Vancouver, BC, V6T 1Z2, Canada

Preface

There is money to be made in the financial industry. Academics, under pressure to exhibit relevance, are happy to point to their consultancies in the City as evidence of their value in the market, and the industry has shown a notable ability to recruit the brightest and best from our Universities.

These observations should not obscure the profound scientific challenges posed by the area of finance. The area has both stimulated and benefited from advances in a range of mathematical sciences, most obviously probability, differential equations, optimization, statistics and numerical analysis. One thinks, for example, of Bernoulli's resolution, in the 18th century, of the St Petersburg Problem through his introduction of a logarithmic utility, of Bachelier's description, at the turn of this century, of the stochastic process we now call brownian motion, of Kendall's investigation, forty years ago, of the statistical unpredictability of stock prices, and of the current enormously fertile interaction between economics and mathematics centred around martingale representations. Looking to the future, some of the mathematical ideas originally motivated by statistical mechanics, and since used to model the large-scale telecommunication networks upon which the financial industry relies, may also provide insight into the very difficult problems that arise in economics concerning interacting systems of rational agents.

This combination of intellectual challenge and practical application, together with the distinction of our speakers, helped to explain the record attendance at this Discussion Meeting. The balanced representation from academia and industry was notable, and contributed to a vigorous, often informal, exchange of ideas.

F. P. KELLY

1

Influence of mathematical models in finance on practice: past, present and future

ROBERT C. MERTON

The mathematics of finance contains some of the most beautiful applications of probability and optimization theory. Yet despite its seemingly abstruse mathematics, finance theory over the last two decades has found its way into the mainstream of finance practice. Today much of the applied financial research on the use of mathematical models takes place within financial institutions. It was not always thus. The scientific breakthroughs in financial modelling both shaped and were shaped by the extraordinary flow of financial innovation which coincided with revolutionary changes in the structure of world financial markets and institutions during the past two decades. The paper covers that development with a focus on the future role of mathematical models in finance practice.

---

1. Introduction

The core of finance theory is the study of the behaviour of economic agents in allocating and deploying their resources, both spatially and across time, in an uncertain environment. Time and uncertainty are the central elements that influence financial behaviour. The complexity of their interaction brings intrinsic excitement to the study of finance as it often requires sophisticated analytical tools to capture the effects of this interaction. Indeed, the mathematical models of modern finance contain some of the most beautiful applications of probability and optimization theory. But of course, all that is beautiful in science need not also be practical; and surely, not all that is practical in science is beautiful. Here we have both. With all their seemingly abstruse mathematical complexity, the models of finance theory have nevertheless had a direct and significant influence on finance practice. Although not unique, this conjoining of intrinsic intellectual interest with extrinsic application is a prevailing theme of research in modern finance.

Although I intend to exemplify this theme, my remarks will not focus on deriving new practical applications of financial models, nor is it my aim to introduce any mathematical tools heretofore unknown to finance which might help to break new ground in the theory. In these directions, the practitioner, the general scientist and the mathematical finance specialist will happily find their cups abundantly filled by the multifarious papers and discussion in this Volume. Instead, in this salutatory paper, I try my hand at providing a frame of reference for what is to follow by describing something of the interaction between mathematical models and finance practice; in the past, in the present and, most importantly, in the impending future.

The origins of much of the mathematics in the papers set forth in the meeting can be traced to Louis Bachelier's (1900) magnificent dissertation on the theory of

Mathematical Models in Finance. Edited by S.D. Howison, F.P. Kelly and P. Wilmott.
Published in 1995 by Chapman & Hall, London. ISBN 0 412 63070 2

speculation. (See Cootner (1964) for an English translation by A. J. Boness. Mandelbrot (1989) provides a brief biography.) Completed at the Sorbonne in 1900, this work marks the twin births of both the continuous-time mathematics of stochastic processes and the continuous-time economics of option pricing. In analysing the problem of option pricing, Bachelier provides two different derivations of the Fourier partial differential equation as the equation for the probability density of what is now known as a Wiener process/brownian motion. In one of the derivations, he writes what is now commonly called the Chapman–Kolmogorov convolution probability integral, which is surely among the earlier appearances of that integral in print. In the other derivation, he takes the limit of a discrete-time binomial process to derive the continuous-time transition probabilities. (This same approach using the binomial process is now applied as a numerical approximation method to solve complicated derivative-security pricing problems; see Merton (1992*b*) for references.) Along the way, Bachelier also developed essentially the method of images (reflection) to solve for the probability function of a diffusion process with an absorbing barrier. This all took place five years before Einstein's discovery of these same equations in his famous mathematical theory of brownian motion. Not a bad performance for a thesis on which the first reader, Henri Poincaré, gave less than a top mark. (Bernstein (1992, p. 19) reports that the thesis was awarded 'mention honorable' instead of 'mention trés honorable'. See Dimand (1993).)

Just as in Bachelier's option-pricing model of 1900, so the normal and binomial distributions are central to today's mathematical models of finance. Although probably coincidental, the choice of subject matter makes it altogether fitting that this Discussion Meeting be held this very week, for the historical precedent took place almost exactly 260 years ago on 12 November 1733. Abraham de Moivre, a long-time Fellow of the Royal Society, issued a short pamphlet in Latin containing the first derivation of what we now call the normal (or gaussian) distribution, as a limit to the binomial process. (As de Moivre (1738) later described the pamphlet in its English translation, 'printed November 12 1733, communicated to some Friends, but never yet made public, reserving to myself the right of enlarging my own thoughts, as occasion shall require.' Stephen Stigler, University of Chicago (personal communication). Stigler also notes that de Moivre published the first version of his *Doctrine of Chances* in *Phil. Trans. R. Soc. Lond.* **27**, 213–264 (1711).)

2. Mathematical models in practice: 1960s and before

Sophisticated mathematical models and a strong influence on practice were not always hallmarks of finance theory. Indeed, Bachelier's work was unknown in the finance literature for more than half a century. (The rediscovery of his work by economists in the early 1950s is generally credited to P. A. Samuelson (1965*a*) via the statistician, L. J. Savage.) During most of this period, finance was almost entirely a descriptive discipline with a focus on institutional and legal matters. Finance theory was little more than a collection of anecdotes, rules of thumb, and shufflings of accounting data. Mathematical models of finance were focused on the time value of money and the most sophisticated tool of analysis was present value. Application of these models by practitioners in non-financial firms was largely confined to staff who set guidelines for capital-budgeting decisions. Although the mathematical modelling of bond-price sensitivity to interest rates (*duration*) had been developed by Macaulay in 1938, there was little evidence of its use in practice more than 20 years later, even by issuer and trader specialists in the debt markets. (Unlike in finance at this time, the mathematical models of the actuarial discipline addressed both the time value of money and the evaluation of uncertainty. The term 'actuary' appears to have been used first as the title for the chief executive of the Equitable Life Assurance Society at its founding in 1762. Over the next 200 years, applications of actuarial modelling involve almost exclusively evaluating mortality risks for life insurance and annuities and casualty risks for non-financial assets. For an overview of the use of actuarial methods in finance, see O'Brien 1992.)

Modern finance begins only in the late 1950s and 1960s. There were important breakthroughs in corporate finance, but the major strides in mathematical modelling came in the area of investments and capital markets. The Markowitz (1952, 1959) mean variance theory of portfolio selection provided a tractable model for quantifying the risk-return tradeoff for general assets with correlated returns. Building on Markowitz's fundamental work, Sharpe (1964) and Lintner (1965) investigated the equilibrium structure of asset prices, and their Capital Asset Pricing Model (CAPM) became the foundational quantitative model for measuring the risk of a security. The CAPM would later form the basis for developing an entire industry to measure the investment performance of professional money managers.

Another important influence of 1960s research on investment practice was the Samuelson (1965*b*)–Fama (1965) efficient markets hypothesis which holds that in a well-functioning and informed capital market, asset-price dynamics are described by a (sub)martingale in which the best estimate of an asset's future price is the current price, adjusted for a 'fair' expected rate of return. Under this hypothesis, attempts to use past price data or publicly available forecasts about future economic fundamentals to predict future security prices are doomed to failure. Earlier, Maurice Kendall (1953) of Royal Statistical Society fame had presented empirical evidence that stock and other speculative price changes were not forecastable. Indeed, Bachelier (1900) had formulated his theoretical option-pricing model assuming a martingale-like behaviour for security prices. It was, however, the Fisher & Lorie (1964) study of historical stock returns that probably first drew practitioner attention to the efficient market hypothesis. Using the newly created data base of the Chicago Center for Research in Security Prices, Fisher & Lorie showed that a randomly selected stock held from the mid-1920s to the mid-1960s would have earned, on average, a 9.4% annual compound return. Returns of this magnitude were believed to be considerably larger than those most professional managers earned for their clients during that period. Rigorous scientific confirmation of this belief was provided by a host of empirical performance studies along lines set by Jensen (1968) who used the CAPM as a benchmark to test for superior performance among United States mutual funds in the postwar period.

3. Mathematical models in practice: 1970s and 1980s

In the late 1960s and early 1970s, models of finance being developed in academe became considerably more sophisticated, involving both the intertemporal and uncertainty aspects of valuation and optimal financial decision making. Dynamic portfolio theory extended and enriched the static Markowitz mean variance model. Intertemporal and international capital asset pricing models expanded the single risk measure, beta, of the Sharpe–Lintner CAPM to multi-dimensional measures of a security's risk. (For references and analysis on dynamic portfolio theory and intertemporal capital asset pricing, see Merton (1992*b*). Solnik (1974) was the first to adopt these models into an international framework. Although static in its formal development, the arbitrage pricing theory of Ross (1976) also provides for multiple dimensions in the measure of a security's risk.) The mathematical tools used in these models – stochastic differential and integral equations, stochastic dynamic programming, and partial differential equations – were a quantum level more complex than had been used previously.

The most important development in terms of impact on practice was the Black–Scholes model for option pricing (Black & Scholes 1973). (For the story of how they developed their model see Bernstein (1992, ch. 11); for extensive references and extensions of the model see Merton (1992*b*).) Virtually from the day it was published, this work brought the field to closure on the subject. The Chicago Board Options Exchange (CBOE) began trading the first listed options in the United States in April 1973, a month before the official publication of the Black–Scholes model. By 1975, traders on the CBOE were using the model to both price and hedge their options positions. Indeed, Texas Instruments created a hand-held calculator which was specially programmed to produce Black–Scholes option prices and hedge ratios. Such a complete and rapid adoption of finance theory into finance practice was unprecedented, especially for a mathematical model developed entirely in theory. That rapid adoption was all the more surprising, as the mathematics used in the model were not part of the standard mathematical training of either academic or practitioner economists.

The basic insight underlying the Black–Scholes model is that a dynamic portfolio trading strategy in the stock can be found which will replicate the returns from an option on that stock. Hence, to avoid arbitrage opportunities, the option price must always equal the value of this replicating portfolio. The resulting pricing formula has only one input which is not directly observable: the volatility of the return on the stock. Just as the CAPM stimulated the development of estimation techniques to measure beta, so the widespread application of the Black–Scholes model created a need for techniques to estimate standard deviations of security returns. In turn, the demand for such estimates in practice created substantial business opportunities and indeed, practitioner-developed models for estimating volatility are among the most sophisticated ones now available.

Black and Scholes, along with others, recognized that their replicating-portfolio approach could be applied to the pricing of general derivative securities with arbitrary nonlinear pay-offs contingent on one or more traded security prices. Hence, at the same time that their work was closing gates on fundamental research on options, it was simultaneously opening new gates by setting the foundation for a new branch of finance called contingent-claims analysis (CCA). The applications of CCA range from the pricing of complex financial securities to the evaluation of corporate capital budgeting and strategic decisions and include, for instance, a unified theory for pricing corporate liabilities and the evaluation of loan guarantees and deposit insurance. Indeed, the theory and mathematical modelling of CCA for these applications have become even more important to finance practice than the original options applications.

The applications of mathematical models in the 1970s had been primarily in equity markets and equity derivative securities. The big, new applications of the 1980s were in the fixed-income arena. The models incorporated major multivariate extensions of the CCA methodology to price and hedge virtually every kind of derivative instrument, whether contingent on equities, fixed-income securities, currencies, or commodities. Dynamic models of interest rates were combined with CCA models to price both cash-market and derivative securities simultaneously. The enormous U.S. national mortgage market could not have functioned effectively without mathematical models for pricing and hedging mortgages and mortgage-backed securities whose valuations are especially complex because of pre-payment options. By the late 1980s, the time-lag for adoption of finance theory models into practice was essentially nil. Indeed, the mathematical models used in finance practice became as sophisticated as any found in academic financial research.

The users of mathematical financial models in the 1970s were U.S. institutional equity investors, market makers and brokers trading U.S. equity options, currency traders, and a few fixed-income traders. During the 1980s, the user base of these models expanded greatly, becoming global in scope and including commercial and investment banks and institutional investors of all types (see Grundfest 1990). Practitioners in financial institutions actually took on a major role in applied research, including the creation of proprietary data bases, development of new numerical methods for solving partial differential equations, and implementation of sophisticated estimation techniques for measuring model parameters.

There are several explanations for the vastly different adoption rates of mathematical models into finance practice during the 1960s, 1970s, and 1980s. Money and capital markets in the United States exhibited historically low volatility in the 1960s: the stock market rose steadily, interest rates were relatively stable, and exchange rates were fixed. Such 'simple' market environments provided little incentive for investors to adopt new financial technology. In sharp contrast, the 1970s experienced several events that led both to structural changes and increasing volatility in financial markets. The more important of these were: the shift from fixed to floating currency exchange rates with the collapse of the Bretton Woods Agreement and the devaluation of the dollar; the world oil price crisis resulting from the creation of the Middle East cartel; the decline in the U.S. stock market in 1973/74, which was larger in real terms than any comparable period in the Great Depression; and the arrival of double-digit inflation and interest rates in the United States. In this environment, the old rules of thumb and simple regression models based on extrapolation of historical relations were wholly inadequate for making investment decisions and managing risk exposure.

During the 1970s, derivative-security exchanges were created to trade listed options on stocks, futures on major currencies, and futures on U.S. Treasury bills and bonds. The success of these markets measured in terms of trading volume can be attributed in good part to the increased demand for managing risks in the volatile economic environment. This success in turn strongly affected the speed of adoption of quantitative financial models. For example, experienced traders in the preceding over-the-counter (OTC) dealer market had achieved a degree of success by using heuristic rules for valuing options and judging risk exposures. However, these rules of thumb were soon to be found to be inadequate for trading in the fast-paced exchange-listed options market with its smaller price spreads, larger trading volume and requirements for rapid trading decisions while monitoring prices in both the stock and option markets. In contrast, formal mathematical models along the lines of the Black–Scholes model were ideally suited for application in this new trading environment.

The growth in sophistication about mathematical models and their adoption into finance practice during the 1980s far exceeded the 1970s and ran in parallel with the extraordinary growth of financial innovation (Finnerty 1992; Miller 1986, 1992). A wave of deregulation in the financial sector as well as huge government budget deficits, especially in the U.S., which increased several times the amount of sovereign debt worldwide that required intermediation, and placement were important factors driving innovation (see Miller 1986, 1992; Merton 1992*a*). They were not, however, the only ones.

Conceptual breakthroughs in finance theory in the 1980s were perhaps fewer and less fundamental than in the 1960s and 1970s, but the research resources devoted to the development and refinement of mathematical models and financial data bases to support them were considerably larger. Moreover, the opportunities and feasibility of implementing these models in practice were also much greater. (See Eckl *et al.* (1990), Marshall & Bansal (1992), and Smith & Smithson (1990) on the development of financial engineering in this period.) Major developments in computing and telecommunications technologies (including the personal computer and the orders-of-magnitude increases in computer speed and memory size) made possible the formation of many new financial markets and substantial expansions in the size of existing ones. Those same technologies made feasible the numerical solution of new complex CCA models with multivariate partial differential equations. They also speeded up the solution of existing CCA models to allow virtually real-time calculations of prices and hedge ratios. CCA and related topics were widely incorporated in top business school curricula during the late 1970s and early 1980s. As a result, by the middle 1980s there was a significant pool of MBA and Ph.D. professionals trained in modern finance theory who were available to put the theory into practice. The pool was further augmented by mathematicians and scientists with advanced degrees attracted to the financial services industry by high salaries and challenging problems.

Success of the new trading markets and intermediated products can itself lead to further success though a process called the 'financial-innovation spiral' (see Merton 1992*a*, *b*, 1993) that proceeds as follows: The proliferation of new trading markets in standardized securities such as futures, options, and swaps makes possible the creation of a wide range of new financial products, many custom designed and sold OTC by financial intermediaries to meet selected needs of investors and corporate issuers. Next, volume in the new markets expands further as the intermediaries themselves trade simply to hedge their own exposures from the products they sold. Such increased volume in turn reduces marginal transactions costs and thereby makes possible the further implementation of new products and trading strategies and this, in turn, leads to still more volume. New markets also evolve as some successful products become standardized and their source of distribution moves from intermediaries to markets. Success of these trading markets and custom products then encourages investment in creating additional markets and products, and so on it goes, spiralling toward the theoretically limiting case of complete markets and zero marginal transactions costs.

The reduction in transactions costs for financial institutions was substantial during the 1980s. The costs of implementing financial strategies for institutions using derivative securities such as futures or swaps are often one tenth to one twentieth the cost of using the underlying cash-market securities. This is especially the case for investments by foreign institutions which are often subject to withholding taxes on either interest or dividends. (See Perold (1992) for an excellent description of the comparative costs of various alternatives for executing basic investment strategies. Although institutions have experienced the greatest reduction in the cost of direct market transactions, retail investors have in places had similar experiences: discount retail brokers in the U.S. now charge as little as 3–5 cents a share commissions on stock transactions.)

The decline in costs does not derive only from reductions in bid–ask spreads and commissions. There are also cost savings from movement down the learning curve. With the cumulative experience of having built several new markets, innovators become increasingly more efficient and the marginal cost of creating additional markets falls.

The same learning-curve effect applies to the application of mathematical models. Beginning in the late 1980s and continuing to the current time, the volume of derivative securities business has shifted substantially from exchange-traded derivatives to more-customized OTC contracts issued by financial institutions directly to their customers. I believe that this shift reflects a growing confidence by institutions in their valuation models that comes not only from technical improvements in the models but also from greater experience in their use. When a firm operates in exchange-traded derivatives, it has current and historical prices to calibrate its valuation model. Competition among transactors in a centralized market provides at least some protection against losses from errors in any one transactor's valuations. In a bilateral OTC transaction, the firm must rely on its valuation models without the benefit of market verification. Hence, without sufficiently reliable mathematical models for valuation, much of the financial innovation that originates in the OTC market could not take place. At the same time, such innovation increases the demand for more sophisticated models. Thus, we see that mathematical modelling both shapes and is shaped by the flow of financial innovation.

4. Mathematical models in practice: 1990s and beyond

Some see the extraordinary growth in derivative securities over the past five years as only a fad. However, a more likely explanation is the vast savings in transactions costs from their use. Looking to the future, with such cost savings we are not going back: derivatives are a permanent part of the mainstream global financial system.

It may be difficult to believe that the pace of general financial innovation during the past decade can sustain itself into the future. However, there are reasons to believe that it can. The decision to implement an innovation involves a trade-off between its benefit and the cost. With secularly declining transactions and learning-curve costs, the threshold benefit required to warrant implementation also declines secularly. Hence, holding fixed the same pace of change in the underlying economic fundamentals as in the past, the implementation of financial innovation is likely to be more rapid since the threshold for change is lower.

With much lower costs of change, it becomes profitable not only to introduce new products and create new markets, but also to change entire institutional arrangements (including geographical and political locations) in response to much smaller shifts in customer tastes or operating costs than in the past. Lower transactions costs, together with the prospect of greater global competition in financial services, form the basis for forecasting substantial increases in both the frequency and the magnitude of institutional changes for both private sector and government financial intermediaries and for regulatory bodies. In the past, mathematical models played a key role in supporting the creation of new products and markets. In the future, that role will expand to include supporting the creation of entire new institutions.

A successful conceptual framework for analysing issues involving the global financial system in the future must address endogenously differences in institutional structure across geopolitical boundaries and in the dynamics of institutional change. The neoclassical-economics perspective addresses the dynamics of prices and quantities. But it is largely an 'institution-free' perspective in which only functions matter. It thus has nothing to say directly about cross-sectional or intertemporal differences in the institutions that serve these functions. In contrast, there is the institutional perspective in which institutions not only matter but are the conceptual 'anchor'. This perspective takes as given the existing institutional structure and views the objective of public policy as helping the institutions currently in place to survive and flourish. Framed in terms of the banks or the insurance companies, managerial objectives are similarly posed in terms of what can be done to make those institutions perform their particular financial services more efficiently and profitably. The institutional perspective addresses cross-sectional differences across borders but is static in focus. Hence, institutional change is exogenous within this perspective.

Drawing on both these perspectives, the functional perspective (see Merton 1992*a*, 1993) takes as given the economic functions performed by financial institutions and seeks to discover what is the best institutional structure for performing those functions at a given time and a given place. It does not posit that existing institutions, whether private sector or governmental, operating or regulatory, will be preserved. Functions thus serve as the conceptual anchor here. Because institutions matter but are not the anchors, institutional changes are endogenous within this perspective.

To illustrate the differences between the institutional and functional perspectives, consider a hypothetical example involving municipal-bond insurance (Merton 1993, pp. 28–29) In the U.S., there are specialized insurance companies that sell insurance contracts which guarantee interest and principal payments on municipal bonds against default by the issuer. The policies are typically sold to the issuer who attaches them to the bonds to give them the highest (AAA) credit rating. To succeed as a guarantor, the insurance company itself must be seen as a very strong credit. In evaluating the firm's competitive standing, a manager with an institutional perspective would focus on other insurance companies as competitors. A manager with a functional perspective would instead focus on the best institutional structure to perform the function, which may not be an insurance company. Consider as one alternative an option exchange which creates a market for put options on municipal bonds. Investors could then achieve the same protection against loss by buying an 'uninsured' municipal bond and a put option on that bond. (With a standard fixed exercise price, the put would actually provide more protection because it covers losses in the value of the bond for any reason, not just issuer default. However, the coverage could effectively be narrowed to only default risk by making the exercise price float to equal the current price of an AAA bond with terms comparable with those of the covered bond.) Note that both structures serve the same function for investors: protection against loss from default. However, the institutions are entirely different: an options exchange is not an insurance company. Furthermore, the put option traded on the exchange is a different product from the insurance guarantee. Nevertheless, although the products and institutions that provide them are both quite different, the economic function they serve is the same.

In certain environments, it is surely possible that an options exchange with market-to-market collateral and a clearing corporation could be a 'better credit' than an insurance company and thereby, also be a superior institutional structure to serve the guarantee function. In such environments, the institutionally oriented manager may miss recognizing the firm's prime competitor. Regulatory bodies for financial services are almost exclusively organized along institutional lines and therefore, they face similar problems. Because options are not insurance products and exchanges are not insurance companies, insurance regulators would have no control over the option exchange even though its product is a perfect substitute for an insurance product.

The increasing flexibility and global mobility of financial institutions, together with the technology for creating custom financial contracts at low cost, have far-reaching implications not only for the regulation of financial services, but for national monetary and fiscal policies as well. Thus, policy-makers are effectively speculating against a long-run trend of declining transactions costs if they assume that traditional frictions within their individual financial systems will continue to allow national governments to pursue monetary and related financial policies with the same degree of control as in the past. If, for instance, the principal mechanism of central bank influence on macroinvestment policy comes from controlling banks that ration credit to their customers (Stiglitz & Weiss 1981), then what would be the impact on that mechanism from the creation of a national mid-market lending market similar to the US mortgage market (Cushman 1993)? Much the same point applies to a nation's fiscal policy which will surely be further constrained not just with respect to taxes targeted on financial services and transactions, but even with respect to general personal and corporate income taxes. (See Lindgren & Westlund (1991) and Umlauf (1993) on the recent Swedish experience with a transaction tax.) Scholes & Wolfson (1992) provide a general development of the theory and application of financial instruments and alternative institutional designs to respond to differing tax and regulatory structures. These techniques have a greatly magnified influence in a low-transaction-cost and global environment.

Risk management is perhaps the central topic for the management of financial institutions in the 1990s. (The 1993 survey on international banking in the *Economist* is devoted entirely to risk management of banks and its implications for bank managers and bank regulators in the future (Freeman 1993). Bankers Trust has been identified as perhaps the prime instance of a major global financial institution that has built its corporate strategy and operations around a mathematically sophisticated risk-management system (see Loomis 1992; Picker 1992; Sanford 1993).) Of course, risk management has always been important to such institutions. However, the focus in the past has been on capital adequacy to ensure performance. Equity capital is a robust all-purpose cushion against unanticipated losses, though at times a quite expensive and inefficient means for doing so. With the vast array of financial instruments and quantitive models for estimating exposures to risk, there is now a greater opportunity to eliminate risk exposures of the firm on a more targeted and efficient basis by hedging specific, non-value-enhancing risks. The cost is that the user of hedging techniques must have a more precise, quantitative assessment of the firm's business risks than the user of equity capital. In turn, the greater need for precision places greater demands on the use and accuracy of mathematical models that measure exposures. With the spread of non-financial firms' use of the hedging alternative to equity capital for managing risk, further growth in the application of these mathematical models is likely in the future.

Institutional investment managers are generally classified by categories in terms of the assets managed: e.g. 'U.S. equities' or 'U.S. corporate debt'. However, with the development of low-cost derivative contracts, these classifications are rapidly losing any meaning. For example, a superior-performing manager in the fixed-income area can be transformed into a superior-performing manager of equities by the use of a swap contract. Suppose a bond manager can outperform standard bond-index returns by 2% (200 basis points) on average. Suppose, however, that an institutional investor is looking for a superior equity manager. If the bond manager places the investor's funds in the bond portfolio and simultaneously, enters into a swap agreement with a third party to pay the returns on the standard bond index and receive the returns on the standard equity index (for instance, in the U.S., the Standard & Poor's 500; up to transaction fees, by arbitrage, the cost of this swap is zero), then the (expected) net return to the investor is (bond index +2%) − bond index + stock index = stock index + 2%. Thus, the superior bond manager is now a superior equity manager. This example shows the growing arbitrariness of traditional institutional categories for investment managers. It also illustrates a more general point, namely, the economic distinctions between institutionally defined debt and equity are becoming increasingly blurred.

This same blurring also occurs among other asset classes. The Bank for International Standards (BIS) international capital requirements for banks set a schedule of capital based on asset-class risk. For instance, US Treasury bonds are in a class that requires zero capital and mortgages are in a class that requires 4% capital. A bank that invests in US Treasury bonds and enters into an amortizing swap in which it pays the total return on those bonds and receives the total payments on mortgages will earn the economic equivalent of holding mortgages directly, but will face a BIS capital requirement much closer to zero than 4%.

All of this suggests that the future will require major revisions in the accounting conventions used in contract enforcement and implementation of regulations. Although it is too early to know what form these revisions will take, I conjecture that the new categories will be defined in terms of equivalent exposures, very much like the 'deltas' of contingent-claim securities in CCA. If this is the case, mathematical models of finance will have a new and enormous area of application. (See Sanford (1993) for a practising Chief Executive Officer's vision of the future financial marketplace and his characterization of the new classification categories in terms of 'particle finance'.)

In summary, in the vast bulk of the past, mathematical models have had a limited and ancillary impact on finance practice. But during the last two decades, these models have become central to practitioners in financial institutions and markets around the world. In the future, mathematical models are likely to have an indispensable role in the functioning of the global financial system including regulatory and accounting activities.

Any virtue can become a vice if taken to extreme, and just so with the application of mathematical models in finance practice. I therefore close with an added word of caution about their use. At times the mathematics of the models become too interesting and we lose sight of the models' ultimate purpose. The mathematics of the models are precise, but the models are not, being only approximations to the complex, real world. Their accuracy as a useful approximation to that world varies considerably across time and place. The practitioner should therefore apply the models only tentatively, assessing their limitations carefully in each application.

I thank Zvi Bodie and Myron Scholes for many helpful discussions.

References

Bachelier, L. 1900 Théorie de la spéculations. *Annales de l'Ecole Normale Supérieure*, vol. 3. Paris: Gauthier-Villars.

Bernstein, P. L. 1992 *Capital ideas: the improbable origins of modern Wall Street*. New York: Free Press.

Black, F. & Scholes, M. 1973 The pricing of options and corporate liabilities. *J. Political Econ.* **82**, 637–654.

Cootner, P. H. (ed.) 1964 *The random character of stock market prices*. MIT Press.

Cushman, Jr, J. H. 1993 Investing in loans to businesses. *New York Times*, March 19, Section D.

De Moivre, A. 1738 *Doctrine of chance*, 2nd edn.

Dimand, R. W. 1993 The case of brownian motion: a note on Bachelier's contribution. *Br. J. History Sci.* **26**, 233–234.

Eckl, S., Robinson, J. N. & Thomas, D. C. 1990 *Financial engineering: a handbook of derivative products*. Oxford: Basil Blackwell.

Fama, E. 1965 The behavior of Stock Prices. *J. Business* **38**, 34–105.

Finnerty, J. D. 1992 An overview of corporate securities innovation. *J. appl. Corporate Finance* **4**, 23–39.

Fisher, L. & Lorie, J. 1964 Rates of return on investments in common stocks. *J. Business* **37**, 1–24.

Freeman, A. 1993 A survey of international banking: new tricks to learn. *The Economist* April 10, 1–37.

Grundfest, J. 1990 Internationalization of the world's securities markets: economic causes and regulatory consequences. *J. Financial Services Res.* **4**, 349–378.

Jensen, M. C. 1968 The performance of mutual funds in the period 1945–64. *J. Finance* **23**, 487–616.

Kendall, M. G. 1953 The analysis of time series, part I: prices. *Jl R. statist. Soc.* **96**, 11–25.

Lindgren, R. & Westlund, A. 1990 Transaction costs, trading volume and price volatility on the Stockholm stock exchange. *Skandinaviska Enskilda Banken Q. Rev.* **2**, 30–35.

Lintner, J. 1965 The valuation of risk assets and the selection of risky investments in stock portfolios and capital budgets. *Rev. Econ. Statistics*, **47**, 13–37.

Loomis, C. J. 1992 A whole new way to run a bank. *Fortune* September 7, 76.

Macaulay, F. 1938 *Some theoretical problems suggested by the movement of interest rates, bond yields, and stock prices since 1856*. New York: National Bureau of Economic Research.

Mandelbrot, B. B. 1989 Louis Bachelier. In *The new Palgrave: a dictionary of economics* (ed. J. Eatwell, M. Milgate & P. Newman). London: Macmillan Press Limited.

Markowitz, H. 1952 Portfolio selection. *J. Finance* **7**, 77–91.

Markowitz, H. 1959 *Portfolio selection: efficient diversification of investment*. New York: Wiley.

Marshall, J. F. & Bansal, V. K. 1992 *Financial Engineering*. Needham Heights, MA: Allyn & Bacon.

Merton, R. C. 1989 On the application of the continuous-time theory of finance to financial intermediation and insurance. *Geneva Papers Risk Insurance* **14**, 225–262.

Merton, R. C. 1992*a* Financial innovation and economic performance. *J. appl. Corporate Finance* **4**, 12–22.

Merton, R. C. 1992*b* *Continuous-time finance* (revised edn). Oxford: Basil Blackwell.

Merton, R. C. 1993 Operation and regulation in financial intermediation: a functional perspective. In *Operation and regulation of financial markets* (ed. P. Englund). Stockholm: The Economic Council.

Miller, M. H. 1986 Financial innovation: the last twenty years and the next. *J. Financial Quant. Anal.* **21**, 459–471.

Miller, M. H. 1992 Financial innovation: achievements and prospects. *J. appl. Corporate Finance* **4**, 4–11.

O'Brien, C. 1992 Actuarial methods in finance. In *The new Palgrave dictionary of money and finance* (ed. P. Newman, M. Milgate & J. Eatwell). London: Macmillan Press Limited.

Perold, A. 1992 BEA associates: enhanced equity index funds. Harvard Graduate School of Business (Case #N9-293-024), Boston, MA.

Picker, I. 1992 Bankers trust's amazing risk machine. *Institutional Investor*. August, 29.

Ross, S. A. 1976 Arbitrage theory of capital asset pricing. *J. Econ. Theory*. **13**, 341–360.

Samuelson, P. A. 1965*a* Rational theory of warrant pricing. *Industrial Management Rev.* **6**, 13–31.

Samuelson, P. A. 1965*b* Proof that properly anticipated prices fluctuate randomly. *Industrial Management Rev.* **6**, 41–49.

Sanford, C. 1993 Financial Markets in 2020. In *Proc. Federal Reserve Bank of Kansas City Economic Symp.*

Scholes, M. S. & Wolfson, M. A. 1992 *Taxes and business strategy: a planning approach.* Englewood Cliffs, NJ: Prentice-Hall.

Sharpe, W. F. 1964 Capital asset prices: a theory of market equilibrium under conditions of risk. *J. Finance* **19**, 425–442.

Smith, C. W. & Smithson, C. W. (eds) 1990 *The handbook of financial engineering.* Grand Rapids, MI: Harper Business.

Solnik, B. H. 1974 An equilibrium model of the international capital market. *J. Econ. Theory* **8**, 500–524.

Stiglitz, J. & Weiss, A. 1981 Credit rationing in markets with imperfect information. *Am. Econ. Rev.* **71**, 393–410.

Umlauf, S. 1993 Transaction taxes and the behavior of the Swedish stock market. *J. Financial Econ.* **33**, 227–240.

Discussion

R. V. Simons (*EDS-Scicon Ltd, U.K.*). Rapid change at financial institutions and further erosion of their role in favour of contracts between parties to perform specific functions is fine when the parties actually perform their contracts, but what about counterparty risk?

R. C. Merton. Effective management of contract-default risk is critical to the continuing gains in economic efficiency from financial innovation. Although surely a complex problem, controlling counterparty risk is as surely within the available technological skills and experience of the financial community. Recent evidence that concern about contract default is influencing financial behaviour, includes the significant resources spent by financial institutions (with less than the highest credit rating) to create AAA subsidiaries for performing their contracting business.

A. D. Wilkie (*Watsons, London, U.K.*). It seems that writing options is very similar to the business of insurance. Insurance companies are required to hold additional solvency reserves in respect of their liabilities. Yet banks and market makers writing options appear to be able to do this with, relatively speaking, much smaller solvency requirements. Is this satisfactory?

R. C. MERTON. Writing options is similar to the business of insurance. However, as with traditional insurance, the level of risk exposure varies considerably across different types of options. For example, options written on a listed exchange are secured by mark-to-market collateral and a clearing corporation. The risks of options on stocks and bonds can be hedged by the purchase of the underlying stock or other securities of the firm. The ability to observe continuous prices for the option liabilities and to hedge those liabilities in the market are not typically available for traditional insurance risks; hence, the greater reliance on larger reserves. Although each case needs detailed investigation, in general banks and market makers are not under-reserved in their option-writing activities.

2

Applied mathematics and finance

S.D. HOWISON

My aim is to make some comments of a quite general nature about the relation between applied mathematics and finance, theoretical and practical. I shall begin with a brief description of a case in which 'technology transfer' from a quite different area of mathematics, the Stefan problem, was helpful with a financial problem, namely the Black–Scholes approach to an American option. I then discuss some more general issues about the role of this kind of mathematics in finance and suggest some possible avenues for future progress.

---

1. Stefan problems and American options

The Stefan problem is a model for the melting or solidification of a pure material by heat transfer. Because the solid/liquid interface is *a priori* unknown, we must solve a free boundary problem, which is inherently nonlinear. We see below that the determination of the early exercise boundary for an American derivative product is a version of the Stefan problem.

Suppose that the temperature in a one-dimensional material $-\infty < x < \infty$ is $u(x,t)$. Suppose too that the material changes phase from solid to liquid or vice versa at a (scaled) temperature $u = 0$, and that this process requires the release or uptake of a latent heat which, again by a suitable scaling, may be taken to be 1. Consider a situation in which solid at the melting temperature occupies the region $-\infty < x < s(t)$, with liquid for $s(t) < x < \infty$. In this case there is just one free boundary $x = s(t)$ separating solid from liquid. Then a simple dimensionless model for the evolution of the free boundary is

$$u_t = u_{xx},$$

for $s(t) < x < \infty$, modelling the flow of heat, with

$$u(s(t),t) = 0, \quad -u_x(s(t),t) = \mathrm{d}s/\mathrm{d}t$$

describing the facts that the phase-change temperature is $u = 0$ and that energy is conserved. (We write u_x for $\partial u/\partial x$ and so on.)

The Stefan problem has a vast literature (see Tarzia (1988) for a bibliography containing 2500 papers). Indeed it has acted as an important canonical problem that has both stimulated important theoretical research (for example, the analysis of weak solutions and their relation to classical solutions) and has had considerable practical and numerical implications; a large number of technologies rely on solidification processes. Similar remarks might be made about generalizations of the Stefan model, such as the phase field model (Caginalp 1990) or the alloy problem (Elliott & Ockendon 1982); in another context altogether, the Navier–Stokes equations have likewise focused attention from all areas of applied mathematics.

We now show how to transform the Stefan problem above into a problem that is closely related to the valuation problem for American options. The main obstacle is

Mathematical Models in Finance. Edited by S.D. Howison, F.P. Kelly and P. Wilmott.
Published in 1995 by Chapman & Hall, London. ISBN 0 412 63070 2

the lack of smoothness of the first derivative of u and we overcome this by integrating in time. We define a new variable $v(x,t)$ by the Baiocchi transformation (Elliott & Ockendon 1982)

$$v(x,t) = \int_{s^{-1}(x)}^{t} u(x,\tau)\,\mathrm{d}\tau,$$

where $s^{-1}(x)$ is the inverse of $s(t)$. It is then clear that $v_t = u$. It is straightforward to show that v satisfies

$$v_t = v_{xx} + 1 \quad \text{for} \quad x > s(t), \tag{1}$$

$$v = 0 \quad \text{for} \quad x < s(t), \tag{2}$$

and that

$$v, v_x \quad \text{are continuous.} \tag{3}$$

Furthermore, in the melting case, when $u \geqslant 0$ everywhere, we have the inequality

$$v \geqslant 0. \tag{4}$$

Lastly, the initial data for v, $v(x,0)$, can be calculated from that of u by using (1). Equations (1)–(4) may be combined into the linear complementarity problem:

$$v \geqslant 0, \quad v_t - v_{xx} - 1 \leqslant 0, \tag{5}$$

$$v(v_t - v_{xx} - 1) = 0, \tag{6}$$

with appropriate initial data.

It is a short step from the linear complementarity problem to a variational inequality formulation of the problem. Equations (5) and (6) are multiplied by a test function $\phi(x,t)$ from a suitable space whose most important property is that its members satisfy the constraint (4). After integration by parts, we have

$$\int \frac{\partial v}{\partial t}(\phi - v) + \frac{\partial v}{\partial x}\left(\frac{\partial \phi}{\partial x} - \frac{\partial v}{\partial x}\right)\mathrm{d}x \geqslant 0,$$

for all test functions. (For details, see, for example Elliott & Ockendon (1982).) This weak version of the problem is essentially equivalent to a minimization over the convex set of functions that satisfy the constraint (together with suitable regularity conditions). It lends itself well to proofs of existence and uniqueness and, since it has the great advantage that the free boundary is given implicitly and need not be tracked explicitly, it is well-suited to numerical methods; a good example of the latter is the projected systematic overrelaxation (SOR) method (Cryer 1971).

We now turn to American options. As described in Wilmott *et al.* (1993), these can also be formulated as linear complementarity problems; the general structure is as follows. Suppose that $V(S,t)$ is the value of an option or other contingent claim depending on an underlying asset price S (the framework extends naturally to contingent claims depending on more than one asset) and with pay-off $V_T(S)$. Then the usual Black–Scholes model leads to the linear complementarity problem

$$V(S,t) \geqslant V_T(S), \quad \mathscr{L}_{\mathrm{BS}} V \leqslant 0, \tag{7}$$

$$V \cdot \mathscr{L}_{\mathrm{BS}} V = 0, \tag{8}$$

with $V(S,T) = V_T(S)$ and where V and V_S are continuous. Here $\mathscr{L}_{\mathrm{BS}}$ is the Black–Scholes differential operator associated with the market model used, so that if the interest rate r is constant and S follows the geometric process

$$\mathrm{d}S = \mu S\,\mathrm{d}t + \sigma S\,\mathrm{d}Z, \quad \mathscr{L}_{\mathrm{BS}} = \frac{\partial}{\partial t} + \frac{1}{2}\sigma^2 S^2 \frac{\partial^2}{\partial S^2} + rS\frac{\partial}{\partial S} - r.$$

The inequalities and equality in (7) and (8) can be interpreted financially: the first inequality is the arbitrage constraint that the option value must lie above the pay-off (otherwise early exercise would lead to an arbitrage profit), whereas the second says that the return from the Black–Scholes portfolio of one option and $-V_S$ shares is at most that from a risk-free deposit, and can be less in regions where early exercise is optimal. The equality says that either early exercise is optimal, in which case the option value is equal to that of the pay-off, or the option satisfies the Black–Scholes equation. Finally, the continuity of V and V_S can also be justified using arbitrage arguments related to the exercise strategy of the holder.

It is a straightforward matter to convert the linear complementarity form of the option problem into a variational inequality for the heat equation, using a logarithmic price scale (i.e. set $S = E\,e^x$ where E is a suitable scale for the pay-off function, such as the strike price for a vanilla option) and a scaled version of the option price. We are then in a position to import both theoretical and numerical methods from the corresponding Stefan problem and we can use methods such as the projected SOR algorithm to perform rapid accurate calculations. (It is also interesting to note that the option problem can be thought of as an optimal stopping problem – stopping being equated with exercise – and this route too leads to a Stefan problem; see van Moerbeke (1976) and Duffie (1992).) I do not give further details here (see Dewynne *et al.* 1993 or Wilmott *et al.* 1993), but I wish to stress the effectiveness of the 'technology transfer' at both theoretical and practical levels. It is our contention that many other areas of applied mathematics are ripe for exploitation in this way.

2. The relationship between mathematics and finance

Let us now turn to some more general issues concerning the relationship between mathematics and finance. Let us begin by hypothesizing the existence of a subject called 'mathematical finance', stimulated by increasing technical sophistication in financial markets (with a concomitant need for models and, consequently, analysis and computation). What is the nature of the interaction of this subject with economic theory, economic practice and, in particular, mathematics? What, if any, future does it have? (One reason for wondering whether it has a future as a formal subject is the difficulty posed by the lure of large salaries and problems of confidentiality.)

There is no doubt that the interaction with theoretical and practical economics has been a spectacular success, and we do not pursue these topics further. Turning to the interaction with mathematics, it may be helpful to consider the following questions.

Is mathematical finance purely a collection of mathematical methods, to be applied routinely to standard problems? Or is there, on the other hand, any feedback into mathematics, in the form of new ideas or new problems requiring the development of new mathematics? One of the intentions of this meeting is to show that there is, and perhaps, to indicate where future developments will take place. This aspect is discussed further below.

With which areas of mathematics does the interaction take place? Evidently, the majority of the interaction is with theoretical and applied probability, although activity in the area of differential equations is growing. It is likely that the latter area will become increasingly prominent because it is especially well-suited to asymptotic and numerical methods; this point is exemplified above and in Dewynne *et al.* (this volume).

Given that the differential equations encountered in finance are well in the mainstream of mathematical research and teaching, why are they so little known outside the finance community? Since Black–Scholes, the premier American general applied mathematical review journal, *SIAM Review*, has published just four papers on the subject (Samuelson 1973; Malliaris 1981, 1983; Ellerman 1984), whereas *SIAM Journal on applied Mathematics* has published only one (Dewynne & Wilmott 1994). (Incidentally, SIAM stands for Society for Industrial and Applied Mathematics; finance is surely the world's largest industry and one of the most mathematical.)

Why are there so few undergraduate courses in mathematics departments, despite the central places occupied by brownian motion and the diffusion equation? Finance is a striking and novel application of these ideas and it is a popular career destination for students. One possible reason is that the subject is regarded as 'unsuitable'; another is that the sheer volume of jargon and terminology, both financial and mathematical, that has to be mastered is too much; a third is that an over-rigorous approach to the subject in the existing literature has been off-putting. No doubt some of these reasons will disappear as more finance research is carried out in mathematics departments.

3. Future directions

According to Duffie (1992, pp. xiii, xiv), 'the decade spanning roughly 1969–79 seems like a golden age of dynamic asset pricing theory', whereas 'the decade or so since 1979 has, with relatively few exceptions, been a mopping up operation'. Should we be pessimistic about the outlook for the subject? In this section I suggest some ways in which the subject can develop that may lead to a fruitful interaction with applied mathematics.

One source of new problems is the explosion of 'exotic' contracts on the market; each new version poses a challenge to the mathematical modeller. Fortunately many of the common contracts can be analysed with a framework that generalizes the Black–Scholes analysis (see Wilmott *et al.* 1993) and do not pose substantial difficulties. In the train of these new contracts comes a range of problems in the efficient and rapid numerical calculation of derivative security values, whether in a real-time environment or, as in the area of risk-evaluation, involving very large numbers of contracts. However, the extension to existing concepts is not large, and for new developments we must look further.

Much of the body of finance theory developed to date is essentially linear, characterized by linear partial differential equations or, in the case of American options, which are of course nonlinear because they are free boundary problems, by a linear complementarity framework. A somewhat analogous body of 'linear' theory has been developed and as exhaustively explored for the Stefan problem, but there has also been much recent activity influenced by the problems posed by nonlinear effects such as surface tension effects or models posed in terms of nonlinear diffusion equations. It may be that finance is on the verge of a similar change of emphasis, prompted by the fact that market behaviour still does not exactly agree with the predictions of current models.

On a macroscopic level, the modelling of transaction costs leads immediately to nonlinear problems, for example as shown in this volume by Davis & Clark, Dewynne *et al.* and Pliska & Selby. Beyond this, models of the interaction between market players are only just beginning to emerge; they inevitably lead to nonlinear models as in Föllmer (this volume) and Schonbucher (1993); the latter considers the effect on the asset price of the existence of a large number of traders who follow the Black–Scholes hedging strategy.

Finally, there is much to be done in the detailed modelling of market behaviour. Patterson (1993) says 'I think that many editors and referees of finance journals have shown a studied lack of interest in [the topic of nonlinear dynamics of markets]'. Much of finance theory is built on the efficient market hypothesis yet empirical evidence, for example, Ziemba (this volume), shows that inefficiencies can exist. The scope for mathematical modelling and analysis here is enormous; it ranges from the delicate analysis of financial time series to models of (possibly irrational) investor behaviour in markets that may be inefficient and/or illiquid. There is no doubt that the practical side of the subject, which is after all its *raison d'être*, will drive intense investigation into these matters. It will be very interesting to see to what extent this investigation involves mathematics and mathematicians.

I am grateful to Jeff Dewynne and John Ockendon for many helpful conversations.

References

Caginalp, G. 1990 The dynamics of a conserved phase-field system. *IMA J. appl. Math.* **44**, 77–94.

Cryer, C. W. 1971 The solution of a quadratic programming problem using systematic overrelaxation. *SIAM J. Control* **9**, 385–392.

Dewynne, J. N., Howison, S. D., Rupf, I. & Wilmott, P. 1993 Some mathematical results in the pricing of American options. *Euro. J. appl. Math.* **4**.

Dewynne, J. N. & Wilmott, P. 1994 A note on average rate options with discrete sampling. *SIAM Jl appl. Math.* (In the press.)

Duffie, D. 1992 *Dynamic asset pricing theory*. Princeton.

Ellerman, D. P. 1984 Arbitrage theory: a mathematical introduction. *SIAM Rev.* **26**, 241.

Elliott, C. M. & Ockendon, J. R. 1982 *Weak and variational methods for free and moving boundary problems*. London: Pitman.

Malliaris, A. G. 1981 Martingale methods in financial decision making. *SIAM Rev.* **23**, 434–443.

Malliaris, A. G. 1983 Itô calculus in financial decision making. *SIAM Rev.* **25**, 481–496.

Patterson, D. M. 1993 Book review. *J. Finance*, March.

Samuelson, P. A. 1973 Mathematics of speculative price. *SIAM Rev.* **15**, 1–42.

Schonbucher, P. 1993 M.Sc. thesis, Oxford University, U.K.

Tarzia, D. A. 1988 A bibliography on moving-free boundary problems for the heat-diffusion equation. University of Firenze, Mathematics Department. (Working paper.)

van Morebeke, P. 1976 On optimal stopping and free boundary problems. *Arch. ration. Mech. Analysis* **60**, 101–148.

Wilmott, P., Dewynne, J. N. & Howison, S. D. 1993 *Option Pricing: mathematical models and computation*. Oxford Financial Press.

Discussion

R. Lacey (*Derivative Investment Advisers Ltd, London, U.K.*). Many concepts of financial mathematics have a simple discrete formulation and a rather more difficult continuous one. Traders are well versed in the discrete case, for example the binomial model, but few could derive or solve the continuous time Black–Scholes equation. It might make things easier if new mathematical ideas were presented in the discrete case with the continuous case constructed in an appendix. If current research were formulated discretely, more traders would participate in applied mathematics rather than consume selected parts of applicable mathematics as interpreted by quantitative analysts. This would enable mathematicians and traders to work together in creating new applied financial mathematics.

Rather than taking the intuitively clear discrete case, constructing the continuous formulation and then creating another discrete formulation for calculation, it would seem to make sense to eliminate the last two steps. In some cases the discrete combinatorial analysis is more difficult than the limiting continuous time case.

S. D. HOWISON. There are several reasons for considering the continuous-time approach as preferable. One is that directly derived discrete models are often more difficult to analyse (for example, from the point of view of stability and error analysis). Also, to many people, the differential equations that arise from the continuous-time limit are easier to understand intuitively (after all, they are essentially the diffusion equation). This has even more force when applied to the nonlinear models that will be increasing in importance in the future.

A. D. WILKIE (*Watsons, Reigate, U.K.*) Two other areas worthy of consideration are time series analysis and stable distributions. Although in the short term share prices and other variables may move apparently in a random walk, equivalent to a continuous gaussian diffusion process, for the much longer term it is desirable to use time series modelling to investigate the way in which investment variables, dividend yields, interest rates, and so on, vary jointly in the long run. There is also evidence that the distributions of changes in share prices and other variables are fat-tailed, both in the short and the long run. The only consistent distribution for both timescales is the stable paretian series, of which the normal distribution is one. But the mathematics of the general stable paretian distribution is difficult so it would be useful to learn more about it.

T. LEACH (*Assured Asset Management plc, Isle of Man, U.K.*). The British regulators should be made aware of the use of derivatives as an essential component in obtaining market exposure to minimize transaction costs, build in flexibility, and separate market and currency risks of international investments.

M. A. H. DEMPSTER (*University of Essex, U.K.*). The approach to valuing American options based on complementarity problems and their equivalent variational inequalities, due originally to Jaillet *et al.* (*Acta appl. Math.* **21**, 263–289 (1990)), is flexible, powerful and easily extended to nonlinear problems. Partial differential equation methods give an accurate numerical approximation of the entire value surface and hence may be used to price many options simultaneously and to compute trivially all the ‘Greeks’ for risk management through hedging portfolios.

3

Stock price fluctuation as a diffusion in a random environment

HANS FÖLLMER

The fluctuation of stock prices is modelled as a sequence of temporary equilibria on a financial market with different types of agents. I summarize joint work with M. Schweizer on the class of Ornstein–Uhlenbeck processes in a random environment which appears in the diffusion limit. Moreover, it is shown how the random environment may be generated by the interaction of a large set of agents modelled by Markov chains as they appear in the theory of probabilistic cellular automata.

1. Introduction

The price fluctuation of a risky asset in a financial market is usually modelled as the trajectory of a stochastic process $(S_t)_{t\geqslant 0}$ on some underlying probability space $(\Omega, \mathscr{F}, P)$. On the other hand, there have been recent attempts to describe such prices in the context of a deterministic dynamical model exhibiting chaotic behaviour (see, for example, Day & Huang 1990). In this paper we are going to insist on the probabilistic interpretation, but in the final section we will make contact with the deterministic point of view.

The probabilistic approach was initiated by Bachelier (1900) who introduced brownian motion as a model for price fluctuation on the Paris stock market. As prices should stay positive, Samuelson (1964) suggested using instead geometric brownian motion defined as the pathwise solution

$$S_t = S_0 \exp(\sigma W_t + mt) \tag{1.1}$$

of the linear stochastic differential equation

$$\mathrm{d}S_t = \sigma S_t \,\mathrm{d}W_t + \mu S_t \,\mathrm{d}t, \tag{1.2}$$

where $m = \mu - \frac{1}{2}\sigma^2$ and (W_t) denotes a standard brownian motion. Geometric brownian motion is now widely used as a basic reference model. In particular, Black & Scholes (1973) developed their approach to the pricing and hedging of options in the context of this model. The Black–Scholes option pricing formula is based on the insight that, in a complete model such as (1.1), any option can be duplicated by a dynamical hedging strategy; this may be viewed as the financial interpretation of Itô's representation of functionals on Wiener space as stochastic integrals of brownian motion. A hedging strategy introduces a technical demand for the underlying asset. But the justification of the formula relies on the assumption that this technical demand does not disturb the stochastic law of motion of the asset. More precisely, it is implicitly assumed that the strategy induces at most an absolutely continuous change of measure so that there is no change in the volatility

Mathematical Models in Finance. Edited by S.D. Howison, F.P. Kelly and P. Wilmott.
Published in 1995 by Chapman & Hall, London. ISBN 0 412 63070 2

structure of the asset. But if hedging occurs on a large scale then one may start to doubt this assumption. A mathematical analysis of this question calls for a closer look at the microeconomic picture behind a diffusion model such as (1.1). This was the starting point for joint work (Föllmer & Schweizer 1993), which is summarized in §§2 and 3.

As prices are generated by the demand of agents who are active on the financial market for the given asset, a diffusion model such as (1.1) should be explained in terms of the interaction of these agents. Kreps (1982) showed that geometric brownian motion can be justified as a rational expectations equilibrium in a market with highly rational agents who all believe in this model; see also Bick (1987). Here we have much less rational behaviour in mind, and in particular various forms of noise trading and of technical trading. This suggests a distinction between different types of agents. In fact, such distinctions have been a major topic in the recent literature on financial markets. In particular, Black (1986) proposed the distinction between information traders and noise traders and argued that the presence of information traders alone would not explain what we actually observe. This claim can be substantiated in a rather striking manner. In Smith *et al.* (1988), experiments are reported where agents are faced with an investment independent distribution (IID) stream of dividends, and they are provided with full information about the distribution. Thus, all uncertainty is eliminated from the formation of price expectations insofar as these are solely based on dividend forecasts. In a model based on information trading, the resulting price process should become deterministic. But in fact 'fourteen out of twenty-two experiments exhibit price bubbles followed by crashes with respect to intrinsic dividend value'; this is compared with 'a panic of the sort that sometimes occurs in a crowded theatre'. Such observations are in striking contrast to any version of the efficient markets hypothesis. It is a major challenge to develop adequate probabilistic models for these phenomena.

The approach described in §2 tries to provide a framework for such questions. Here we are just taking a first step and we are still far from reaching satisfactory answers. The idea is to view the price process as a sequence of temporary equilibria in a market with agents whose demand may involve liquidity demand, a subjective assessment of adequate price levels and noise trading. Assuming a simple log-linear structure of individual excess demand, logarithmic equilibrium prices are computed as weighted averages of individual reference levels. In the simplest case, a trader would just use the last price as his reference level. On a logarithmic scale, this would lead to a random walk for asset prices, hence to the basic reference model (1.1) of geometric brownian motion in the diffusion limit. We then go just one step beyond the random walk case and show how a combination of information trading and a crude simplification of noise trading leads to a class of Ornstein–Uhlenbeck processes in discrete time and in a random environment. The random environment is related to a random fluctuation in the behavioural characteristics of agents, for example in the proportion between different types of agents who are active on the market. With a different interpretation, this class of processes also appears in Orléan & Robin (1991). Models with a similar flavour as to the distinction of different types of agents have been suggested by Black (1986), Day & Huang (1990), De Long *et al.* (1990*a*, *b*), Frankel & Foot (1986) and Kirman (1993), among others. In §3 we pass to the diffusion limit and summarize results of Föllmer & Schweizer (1993) concerning the ergodicity of these processes and the structure of their equilibrium distributions. We obtain a rich class of Pearson-type distributions on a logarithmic scale. In certain cases the invariant distribution may be a continuous but singular mixture of normal distributions; this is related to results of Diaconis & Shahshahani (1986) on the random iteration of affine maps.

Only a crude simplification of technical trading is consistent with the log-linear models in §§2 and 3. In §4 we mention a more detailed analysis of technical trading by Frey & Stremme (1993). Here technical traders are assumed to interact with agents who, if left alone, would generate a geometric brownian motion. The impact of technical trading increases volatility, and this leads to the question of which models have a fixed point property under this transformation.

The final section contains some general remarks on the probabilistic structure of the random environment when the number of agents involved in the formation of price expectations and preference become large. This seems to call for concepts and techniques from the theory of interacting particle systems; see Föllmer (1974), Durlauf (1993) and Brock (1993) for attempts in that direction. In models motivated by statistical mechanics, one has in mind local forms of interaction; an interaction of mean field type is usually viewed as merely a simplifying device to circumvent the deeper problems related to local interaction. But in an economic context, agents are directly affected by macroeconomic signals about aggregate quantities. This information is carried by the empirical distribution of individual agents' characteristics. Thus it makes intrinsic sense to consider interactions with a local and a global component. In a large system where the transition mechanism has the structure of a probabilistic cellular automaton, the evolution on the level of empirical distributions proceeds in a deterministic and possibly chaotic manner. This will affect the formation of reference levels of individual agents and hence the resulting price trajectory. Thus we are making contact with the deterministic point of view mentioned above.

The deterministic evolution of empirical distributions of agents' characteristics would be a key component in the formation of temporary price equilibria. But it would not necessarily be a sufficient statistic. In fact it seems reasonable to view the set of agents who are directly involved in the formation of successive prices as a subset of a much larger set of agents constituting the environment to which traders react. Though randomness of the empirical distributions would disappear in the limit of the larger set, it would still be present in the smaller sample of agents who are active on the market. This is a first reason to retain the probabilistic point of view. But the theory of interacting particle systems suggests a second and more fundamental reason. If in fact interaction appears in the transition itself and not just in the individual probability distributions of the transition, this may cause a phase transition and a corresponding breakdown of the law of large numbers. Thus, randomness may appear on the level of empirical distributions even in the limit of a very large environment.

Of course, it may seem beyond any hope to turn such qualitative features of the random environment into models which are relevant in a quantitative and statistical sense. A more modest purpose is to support, from a mathematical point of view, the attitude of a seasoned practitioner to look beyond the technical analysis of a given price trajectory and to be aware of the random environment producing random shifts in the 'mood' of the market. On a more technical level, a random environment, and in particular the induced randomness of volatility, leads to incompleteness of the financial market. Thus, our discussion emphasizes the need for constructing reasonable hedging strategies in the context of an incomplete model.

2. Asset prices as temporary equilibria

Let A be a finite set of agents who are active on the market for a single speculative asset. In reaction to a proposed stock price p for period k, and depending on circumstances summarized by a sample point ω in some probability space, each agent $a \in A$ forms an excess demand $e_{a,k}(p, \omega)$. The actual stock price $S_k(\omega)$ in period k is then determined by the equilibrium condition

$$\sum_{a \in A} e_{a,k}(S_k(\omega), \omega) = 0. \tag{2.1}$$

Thus, we obtain a sequence of temporary price equilibria viewed as a stochastic process

$$S_k(\omega) \quad (k = 0, 1, \ldots) \tag{2.2}$$

on the underlying probability space $(\Omega, \mathscr{F}, P)$. Let us assume that individual excess demand takes the simple log-linear form

$$e_{a,k}(p, \omega) = \alpha_{a,k} \log(\hat{S}_{a,k}(\omega)/p) + \delta_{a,k}(\omega), \tag{2.3}$$

where $\delta_{a,k}$ may be viewed as a liquidity demand and $\hat{S}_{a,k}$ denotes an individual reference level of agent a for period k. Then the logarithmic equilibrium price in period k is determined via (2.1) as an aggregate,

$$\log S_k(\omega) = \sum_{a \in A_k(\omega)} \bar{\alpha}_{a,k} \log \hat{S}_{a,k}(\omega) + \delta_k(\omega), \tag{2.4}$$

of individual price assessments and liquidity demands.

Example 2.5. Suppose that traders have no sense of the direction where the market is supposed to be going, and that they take the last price S_{k-1} as their reference level $\hat{S}_{a,k}$. Then the logarithmic price process takes the form

$$\log S_k(\omega) = \log S_{k-1}(\omega) + \delta_k(\omega) \quad (k = 0, 1, \ldots) \tag{2.6}$$

of a random walk. Standard assumptions on the sequence (δ_k) of aggregate liquidity demand imply convergence to a brownian motion with drift under suitable rescaling. In this case, the diffusion limit (S_t) of the price process is given by the basic reference model (1.1) of geometric brownian motion.

If individual assessments $\hat{S}_{a,k}$ depend on individual perceptions of fundamental values, or if proposed prices are taken as a signal for a shift in the fundamentals, we must expect that other diffusion models will appear in the limit. With the following assumptions we go just one step beyond the simplest case considered in the previous example.

Consider first an information trader whose reference level $\hat{S}_{a,k}(\omega)$ is based on his current perception $F_{a,k}(\omega)$ of the fundamental value of the asset and on the idea that the next price should move closer to the perceived fundamental value. We assume that it takes the simple form

$$\log \hat{S}_{a,k} = \log S_{k-1} + \beta_{a,k}(\log S_{k-1} - \log F_{a,k}). \tag{2.7}$$

If only such information traders are active on the market, the resulting logarithmic price process behaves like an Ornstein–Uhlenbeck fluctuation around the time-dependent levels of aggregate perceived fundamentals.

In a caricature model of noise trading, we assume that a noise trader takes the proposed price seriously as a signal and simply replaces the fundamental quantity $F_{a,k}$ in (2.7) by p. Thus the excess demand is of the form

$$\log \hat{S}_{a,k} = \log S_{k-1} + \gamma_{a,k}(\log S_{k-1} - \log p). \tag{2.8}$$

A similar structure appears if we use a very crude log-linear approximation of the technical demand induced by hedging strategies based on the Black–Scholes formula (see §4 below). If only such traders are active on the market then the resulting logarithmic price process has the structure of a random walk.

To study the interactive effect of these different types of behaviour, we simply assume that the individual reference level is of the mixed form

$$\log \hat{S}_{a,k} = \log S_{k-1} + \beta_{a,k}(\log S_{k-1} - \log F_{a,k}) + \gamma_{a,k}(\log S_{k-1} - \log p), \tag{2.9}$$

with random coefficients $\beta_{a,k}(\omega)$ and $\gamma_{a,k}(\omega)$. The randomness in these coefficients can be used, for example, to model a random fluctuation in the proportion between different types of agents. The resulting price process takes the form

$$\log S_k = \log S_{k-1} + \tilde{\epsilon}_k(\log S_{k-1} - \log F_k) + \epsilon_k, \tag{2.10}$$

with suitable aggregates $\tilde{\epsilon}_k$, ϵ_k and F_k. Here the random coefficients $\tilde{\epsilon}_k(\omega)$ may become positive if the effect of noise trading becomes too strong. Motivated by the experiments mentioned in the introduction, we concentrate on the effects of the two random sources $(\tilde{\epsilon}_k)$ and (ϵ_k) and leave completely aside the randomness induced by the fundamentals and by their individual perceptions. So we simply assume that the sequence (F_k) is constant, and then our price process takes the form

$$S_k = \exp(X_k + \log F), \quad (k = 0, 1, \ldots) \tag{2.11}$$

where the fluctuation of logarithmic stock prices around the level $\log F$ is given by

$$X_k - X_{k-1} = \tilde{\epsilon}_k X_{k-1} + \epsilon_k, \quad (k = 0, 1, \ldots). \tag{2.12}$$

If the first source of randomness is reduced to a deterministic constant $\tilde{\epsilon}$, then we would have an Ornstein–Uhlenbeck fluctuation, recurrent for $\tilde{\epsilon} < 0$ and transient for $\tilde{\epsilon} > 0$. In the general case (2.12), the logarithmic price process may be viewed as an Ornstein–Uhlenbeck process in a random environment. In periods where $\tilde{\epsilon}_k$ assumes positive values the Ornstein–Uhlenbeck process will change from its usual recurrent behaviour to a highly unstable transient behaviour. Nevertheless, the overall behaviour will remain ergodic if this destabilizing effect is on average not too strong; a precise criterion was given by Brandt (1986). We will now look at a diffusion approximation of (2.12) which will lead to a class of Ornstein–Uhlenbeck diffusions in an ergodic environment. In that context, we will formulate an analogue of Brandt's ergodicity criterion and in the ergodic case we will get a clearer view of the structure of invariant distributions.

3. Ornstein–Uhlenbeck processes in a random environment

Consider a sequence of discrete-time equilibrium price processes $(S_k^n)_{k=0,1,\ldots}$ given by (2.11) and (2.12). If we identify X^n with the continuous-time process $X_t^n := X_{[nt]}^n$ $(0 \leqslant t < \infty)$ then equation (2.12) translates into the stochastic differential equation

$$\mathrm{d}X_t^n = X_{t-}^n \,\mathrm{d}\tilde{Z}_t^n + \mathrm{d}Z_t^n, \tag{3.1}$$

where the processes $\tilde{Z}^n$ and Z^n are defined in terms of partial sums of the sequences $(\tilde{\epsilon}_k^n)$ and (ϵ_k^n). Under mild assumptions on these two sources of randomness, the sequence $(\tilde{Z}^n, Z^n)$ converges in distribution to a continuous semi-martingale $(\tilde{Z}, Z)$. A general convergence theorem of Słomiński (1989) implies that (X^n, S^n) converges in distribution to (X, S) where X is the solution of the stochastic differential equation

$$dX_t = X_t\, d\tilde{Z}_t + dZ_t, \tag{3.2}$$

and

$$S_t = S_0 \exp(X_t - X_0). \tag{3.3}$$

Itô's formula shows that the strong solution of equation (3.2) is given by

$$X_t = \exp(\tilde{Z}_t - \tfrac{1}{2}\langle\tilde{Z}\rangle_t)\left\{X_0 + \int_0^t \exp[-(\tilde{Z}_s - \tfrac{1}{2}\langle\tilde{Z}\rangle_s)]\, d(Z - \langle\tilde{Z}, Z\rangle)_s\right\}, \tag{3.4}$$

where $\langle\tilde{Z}, Z\rangle$ denotes the pathwise covariance process.

Examples 3.5. (*a*) Suppose that the first course of randomness in equation (3.1) vanishes as we pass to the continuous-time limit so that $\tilde{Z} = 0$. Then (3.4) reduces to $X_t = X_0 + Z_t$. Under homogeneity assumptions on the second source of randomness, Z will be a brownian motion,

$$dZ_t = \sigma\, dW_t + m\, dt, \tag{3.6}$$

with constant variance and constant drift, and the price process S is given by the basic reference model (1.1) of geometric brownian motion.

(*b*) If the first source of randomness in equation (3.1) produces in the limit an absolutely continuous drift but no additional noise, i.e. $d\tilde{Z}_t = \tilde{m}_t\, dt$, and if Z is given by (3.6), then the limiting equation (3.2) describes an Ornstein–Uhlenbeck process,

$$dX_t = \tilde{m}_t X_t\, dt + \sigma\, dW_t + m\, dt, \tag{3.7}$$

in a random environment specified by the process $(\tilde{m}_t)$. In the special case where $\tilde{m}_t(\cdot)$ reduces to a constant $\tilde{m} \neq 0$, X is an Ornstein–Uhlenbeck process around the constant level $-m/\tilde{m}$, transient for $\tilde{m} > 0$ and recurrent for $\tilde{m} < 0$. Thus, the price process S is a geometric Ornstein–Uhlenbeck process. In the recurrent case $\tilde{m} < 0$, it may be viewed as a canonical reference model in the class of stationary price processes.

(*c*) Now suppose that the process $(\tilde{m}_t)$ in (3.7) is itself a recurrent Ornstein–Uhlenbeck process around some level $\tilde{m}$ with variance v^2 and drift parameter $\alpha < 0$, defined as the pathwise solution of

$$d\tilde{m}_t = \alpha(\tilde{m}_t - \tilde{m})\, dt + v\, d\tilde{W}_t, \tag{3.8}$$

with respect to a second Wiener process $\tilde{W}$. Thus, the price fluctuation is described as a geometric Ornstein–Uhlenbeck process in an Ornstein–Uhlenbeck random environment. Now consider a passage to the limit where $v \to \infty$ and $\frac{1}{2}v^2/\alpha^2 = \tilde{\sigma}^2$ is fixed. This implies the pathwise convergence

$$\lim_{v\to\infty} \int_0^t \tilde{m}_s(\eta)\, ds = \tilde{\sigma}\tilde{W}_t(\eta) + \tilde{m}t, \tag{3.9}$$

and so equation (3.7) is transformed into equation

$$dX_t = X_t(\tilde{\sigma}\, d\tilde{W}_t + \tilde{m}\, dt) + \sigma\, dW_t + m\, dt. \tag{3.10}$$

Equation (3.10) suggests to consider a random environment of the general form

$$d\tilde{Z}_t = \tilde{m}_t\,dt + \tilde{\sigma}_t\,d\tilde{W}_t, \quad dZ_t = m_t\,dt + \sigma_t\,dW_t, \quad d\langle \tilde{Z}, Z\rangle_t = \gamma_t\,dt, \tag{3.11}$$

where $\tilde{W}$, W are standard brownian motions with correlation (ρ_t) and $\gamma_t = \tilde{\sigma}_t \sigma_t \rho_t$. Adapting the analysis of Brandt (1986) for discrete-time processes of the form (2.12) to the case of continuous time, it can be shown that the price process behaves in a stable recurrent manner if the random environment $(\tilde{m}_t, \tilde{\sigma}_t, m_t, \sigma_t, \gamma_t)$ is ergodic, and that it is on average not too destabilizing.

Theorem 3.12. *If*

$$E[\tilde{m}_0] < \tfrac{1}{2}E[\tilde{\sigma}_0^2] \tag{3.13}$$

then the logarithmic price process X converges to an ergodic process,

$$\lim_{t\uparrow\infty} |X_t - \hat{X}_0 \circ \theta_t| = 0, \quad P-\text{a.s.}, \tag{3.14}$$

where

$$\hat{X}_0 = \int_{-\infty}^{0} \exp\left\{-\int_s^0 [\tilde{\sigma}_u\,d\tilde{W}_u + (\tilde{m}_u - \tfrac{1}{2}\tilde{\sigma}_u^2)\,du]\right\}[\sigma_s\,dW_s + (m_s - \gamma_s)\,ds]. \tag{3.15}$$

Let us now go back to the time homogeneous situation in equation (3.10), where W, $\tilde{W}$ are Wiener processes with constant correlation ρ. In the ergodic case $\tilde{m} < \frac{1}{2}\tilde{\sigma}^2$, the invariant distribution μ of X can be computed explicitly. In fact, we can combine the effect of our two random sources and rewrite (3.10) as the stochastic differential equation

$$dX_t = (\tilde{m}X_t + m)\,dt + \sqrt{(\sigma^2 + 2\rho\sigma\tilde{\sigma}X_t + \tilde{\sigma}^2 X_t^2)}\,dB_t, \tag{3.16}$$

where B is invariant again a standard brownian motion. By Kolmogorov's formula, the density h of the invariant measure μ is given by

$$[\log h(x)]' = -2\frac{[\rho\sigma\tilde{\sigma} - m + (\tilde{\sigma}^2 - \tilde{m})\,x]}{(\sigma^2 + 2\rho\sigma\tilde{\sigma}x + \tilde{\sigma}^2 x^2)}, \tag{3.17}$$

and so μ belongs to the family of Pearson type distributions; see Johnson & Kotz (1970). Integration of (3.17) leads to the following description of the invariant density; here we take the opportunity to correct some errors in the formulation of theorem (5.1) in Föllmer & Schweizer (1993). Note first that for $\tilde{\sigma}^2 = 0$ the condition for ergodicity reduces to $\tilde{m} < 0$, and that μ is the normal distribution with mean $-m/\tilde{m}$ and variance $-\frac{1}{2}\sigma^2/\tilde{m}$. Let us now assume $\tilde{\sigma}^2 > 0$.

Theorem 3.18. *For $|\rho| < 1$ the invariant distribution is given by the density*

$$\text{const}\cdot[\sigma^2(1-\rho^2) + (\tilde{\sigma}x + \rho\sigma)^2]^{-(1-\tilde{m}/\tilde{\sigma}^2)} \exp\left[-\frac{2(\rho\tilde{m}\sigma - m\tilde{\sigma})}{\sigma\tilde{\sigma}^2\sqrt{(1-\rho^2)}}\arctan\frac{\tilde{\sigma}x + \rho\sigma}{\sigma\sqrt{(1-\rho^2)}}\right]; \tag{3.19}$$

for $\rho = m = 0$ this reduces to

$$\text{const}\cdot[1 + (\tilde{\sigma}^2/\sigma^2)\,x^2]^{-(1-\tilde{m}/\tilde{\sigma}^2)}. \tag{3.20}$$

For $|\rho| = 1$ the density is

$$\text{const}\cdot(\tilde{\sigma}x + \rho\sigma)^{-2(1-\tilde{m}/\tilde{\sigma}^2)} \exp[2(\rho\tilde{m}\sigma - m\tilde{\sigma})/\tilde{\sigma}^2(\tilde{\sigma}x + \rho\sigma)]. \tag{3.21}$$

We could consider variations of this setting where the environment changes on a slower time scale. Suppose, for example, that we look at an Ornstein–Uhlenbeck process whose parameter $\tilde{m}$ changes at periods k so that $\tilde{m}_k$ ($k = 1, 2, \ldots$) is an IID sequence with finitely many values and expectation < 0. The invariant distribution is a mixture of normal distributions. The mixing measure is an invariant measure for the random iteration of affine maps corresponding to the induced change of variance, and these maps are on average contracting. We can now apply results of Barnsley & Elton (1988) and of Diaconis & Shahshahani (1986) to obtain information about the invariant measure and, in particular, conditions for the appearance of continuous but singular distributions; see Föllmer & Schweizer (1993) for details.

4. On the effect of hedging strategies

Consider an agent $a \in A$ whose purpose is to replicate a European call option with striking price K and expiration date T. Suppose the agent computes the hedging strategy in terms of the basic reference model of geometric brownian motion for the underlying asset. The resulting demand at time t is given by

$$d_{c,\tau}(x, \eta) = \Phi[(1/\sqrt{s})(x - c + \tfrac{1}{2}s)], \tag{4.1}$$

where Φ denotes the distribution function of the standard normal distribution, $x = \log S_t$, $c = \log K$, $\tau = T - t$, $s = \eta^2 \tau$, and where η is the volatility parameter chosen by the agent at time t under the assumption that it will stay constant from that time on. In reaction to a proposed price p at time $t = t_k$, the resulting excess demand can be approximated by

$$e_{a,k}(p, \omega) = \gamma_{c,\tau}(X_{k-1}, \eta_k)(\log p - X_{k-1}), \tag{4.2}$$

where
$$\gamma_{c,\tau}(x, \eta) = \frac{\partial}{\partial x} d_{c,\tau}(x, \eta) = \frac{1}{\sqrt{(2\pi s)}} \exp\left[-\frac{1}{2s}(x - c + \tfrac{1}{2}s)^2\right]. \tag{4.3}$$

Note that there is a positive feedback effect. A price increase from S_t to $p > S_t$ induces a positive excess demand.

The excess demand in (4.2) is consistent with the simple log-linear structure in §2 only after a crude simplification where the coefficient $\gamma_{c,\tau}(x, \eta)$ is replaced by a constant parameter. For a more careful analysis of the effect of hedging on the underlying price process, Frey & Stremme (1993) have used the following extension of the arguments in §§2 and 3. Consider the simplest case where only two types of agents are active on the market: agents who are involved in hedging, and agents who, if left alone, would generate a logarithmic brownian motion with volatility parameter σ as in example (2.5). The mean excess demand of hedgers resulting at time $t = t_k$ from a common choice η_k of the volatility parameter and from a time-invariant distribution over striking prices and over remaining time intervals is of the form

$$\Gamma(X_{k-1}, \eta_k)(\log p - X_{k-1}). \tag{4.4}$$

For a fixed proportion r of hedging traders, the total mean excess demand becomes

$$r\Gamma(X_{k-1}, \eta_k)(\log p - X_{k-1}) + (1 - r)[\alpha(X_{k-1} - \log p) + \delta_k]. \tag{4.5}$$

Under suitable conditions, Frey & Stremme (1993) show that the rescaled logarithmic price process converges to the unique strong solution of the stochastic differential equation

$$\mathrm{d}X_t = H(X_t, \eta_t)[\sigma\, \mathrm{d}W_t + m\, \mathrm{d}t], \tag{4.6}$$

where
$$H(x,\eta) = (1-r)/[1-r-(r/\alpha)\,\Gamma(x,\eta)]. \tag{4.7}$$

Thus the resulting actual volatility is of the form

$$\sigma^*(x,t) = H(x,\eta_t)\,\sigma. \tag{4.8}$$

Since Γ is positive, the volatility induced by the combined effect of hedging strategies and of reference traders is always greater than the initial reference volatility σ. If all agents involved in hedging use as their volatility estimate the current volatility observed at time t, this leads to the equilibrium equation,

$$\sigma^*(x) = H(x,\sigma^*(x))\,\sigma, \tag{4.9}$$

for the actual volatility.

Simulations reported in Frey & Stremme (1993) indicate that the increase in volatility in the present model, where hedging interacts with the simple trading in example (2.5), is significantly larger than in the rational equilibrium setting studied by Brennan & Schwarz (1989). It is not obvious which kind of modelling comes closer to the realities of the stock market, and so it seems interesting to explore this difference further. Let us also mention the following fixed point problem. Hedging strategies are based on an initial diffusion model; in the above discussion it was taken to be geometric brownian motion. We have seen that the introduction of hedging strategies affects the initial model in a significant way because it changes the volatility structure. It seems therefore reasonable to look for an initial model which, given the information about the distribution of hedgers, has already anticipated this effect.

5. Some comments on the specification of the random environment

In this section we discuss some general features of the source of randomness described by the coefficients $\tilde{\epsilon}_k$ in our simple model (2.12) for the fluctuation of logarithmic stock prices. The randomness in these coefficients could be generated by a random fluctuation in the proportion of information traders among agents who are active on the market. If this is modelled by a simple Markov chain of the Ehrenfest urn model type, the diffusion limit takes the form (3.7), where $(\tilde{m}_t)$ is a recurrent Ornstein–Uhlenbeck process around some equilibrium level $\tilde{m} < 0$. In periods where $\tilde{m}_t$ deviates to a positive level, the Ornstein–Uhlenbeck process (X_t) will change from its usual mean reverting to a highly transient behaviour with quite erratic fluctuations. This may be viewed as a simple caricature of the appearance of bubbles and crashes in the context of a model whose overall behaviour is ergodic. Such phenomena becomes more pronounced if the fluctuation of the proportion is modelled by Markov chains with some form of attractive interaction, where the equilibrium distribution becomes U-shaped rather than bell-shaped. A specific model of this type, where equilibria are identified as Beta distributions, was introduced in joint work with A. Kirman; see Kirman (1993).

To see more clearly the effects of interaction in the formation of agents' reference levels, it seems desirable to bring in concepts and techniques from the theory of interacting particle systems. For interacting random preferences in an Arrow–Debreu type exchange economy, such an attempt has been made in Föllmer (1974); see also Durlauf (1993) and Brock (1993). In our present dynamical context one might proceed as follows. Let us stay in the discrete time setting of §2. Let C denote the set of individual characteristics determining the agent's choice of a reference level. The dynamics of the random environment for the diffusion of stock prices is then described by a Markov chain,

$$P(x, \mathrm{d}y) = \prod_{a \in A} P_a(x, \cdot), \tag{5.1}$$

on the product space $E = C^A$, where $P_a(x, \cdot)$ is a probability distribution on C which depends on the configuration $x = (x_b) \in E$. Such Markov chains may be viewed as probabilistic cellular automata; see, for example Lebowitz *et al.* (1990). Note that the individual transition probabilities P_a have an interactive structure, but that the transition itself is made independently by different agents. For the individual law $P_a(x, \cdot)$ we could assume local dependence on some set of 'neighbours' $b \in N(a)$. But in contrast to models in statistical physics, it makes immediate sense to introduce at the same time a dependence on macroeconomic information incorporated in the empirical distribution,

$$\rho_A(x, \cdot) = \frac{1}{|A|} \sum_{a \in A} \delta_{x_a}, \tag{5.2}$$

of individual characteristics. In the simplest case of a pure mean-field interaction, and without distinction of different types of agents, we would assume

$$P_a(x, \cdot) = Q(x_a, \rho_A(x, \cdot), \cdot), \tag{5.3}$$

where Q is a probability kernel from $C \times \mathcal{M}_1(C)$ to C and $\mathcal{M}_1(C)$ denotes the set of probability measures on C.

In the limit of an infinite set A of agents, Kolmogorov's 0–1 law for the product measure $P(x, \cdot)$ implies a deterministic evolution on the level of empirical distributions. More precisely, let us define the set

$$E_1 := \{x \in E \mid \exists \rho(x, \cdot) = \lim \rho_{A_n}(x, \cdot)\}, \tag{5.4}$$

for a suitable sequence $A_n \uparrow A$. In case (5.3) the law of large numbers shows that, for any $x \in E_1$, the measure $P(x, \cdot)$ is concentrated on the set E_1, and that the empirical distribution satisfies

$$\rho(y, \cdot) = \int Q(z, \rho(x, \cdot), \cdot)\, \rho(x, \mathrm{d}z) = F(\rho(x, \cdot)), \tag{5.5}$$

for $P(x, \cdot)$ almost all $y \in E$. Thus, the evolution of the random environment is specified by the deterministic, and possibly chaotic, iteration of the map F acting on empirical distributions.

In the presence of both a local and a macroeconomic component in the interaction, the analysis becomes more involved. Only in special cases can we obtain a simple macroscopic equation for the deterministic evolution of the empirical distribution. As an illustration of the interplay between individual behaviour and the empirical distribution, we consider the following simple voter model with $C = \{0, 1\}$. The individual transition probability is described as a convex combination,

$$P_a(x, 1) = \alpha p(x_a) + \beta r_a(x) + \gamma r(x), \tag{5.6}$$

where $r_a(x)$ is the proportion of '1' in some finite neighbourhood $N(a)$ and where $r(x) = \rho(x, 1)$ is the proportion of '1' in the whole population. If the graph structure of A is such that averages of $r_a(x)$ converge to $r(x)$, then the evolution of r is given by

$$r(y) = \alpha[r(x)\, p(1) + (1 - r(x))\, p(0)] + (1 - \alpha)\, r(x), \tag{5.7}$$

for $x \in E_1$ and for $P(x, \cdot)$ almost all $y \in E_1$. It is now easy to see that the empirical distributions converge to the equilibrium distribution μ of the Markov chain on C determined by the probabilities $p(0)$, $p(1) \in (0,1)$. Now consider the Markov chain $Q(x, dy)$ on $E = C^A$ obtained if in (5.6) we replace the macroscopic signal $r(x)$ by the constant parameter $\mu(1)$. This process has only local interaction. For $\alpha \neq 0$ a basic convergence theorem of Vassershtein (1969) implies convergence of the process with transition probability Q to a unique equilibrium distribution M on E. Using a perturbation of the Dobrushin–Vassershtein contraction technique, we can now show that the interactive system described by (5.1) and (5.6) converges to the same limit distribution M. A nonlinear dependence of the transition probabilities in (5.7) on $r(x)$ and $r_a(x)$ may, of course, lead to non-uniqueness of the equilibrium distribution and to a much more erratic behaviour on the macroscopic level.

In the dynamical model (5.1), individual transition probabilities are interactive, but the transition itself is not. An interactive structure in the transition would be captured by a model where the measure $P(x, \cdot)$ is not a product measure, but a Gibbs measure with respect to system of conditional probabilities depending on x. For a given agent a, we would thus specify the conditional probability for his next state c, given that the previous configuration was x and that the other agents $b \neq a$ assume states $x_b \in C$. If this additional interaction appearing in the transition itself becomes too strong then there may occur a phase transition i.e. the measure $P(x, \cdot)$ will not be uniquely determined by x. The actual choice of the new distribution $P(x, \cdot)$ constitutes an additional source of randomness. This may be viewed as the endogenous production of a random environment generated by the interaction of agents.

In conclusion, we argue that diffusion models for the price evolution of a financial asset should be viewed in the context of a random environment generated by the interaction of agents. This environment can be a rich source of random and/or deterministic chaotic fluctuations of asset prices. As in the experiments described in Smith *et al.* (1988), this source will be active even in situations where perceptions of the fundamentals are constant and thus do not generate any randomness. At least on a qualitative and conceptual level, the theory of Gibbs measures can contribute to an awareness of this background, and of the resulting incompleteness of the financial market, when applying sophisticated tools of stochastic analysis to the analysis of derivative securities.

References

Bachelier, L. 1900 Théorie de la spéculation. *Ann. Sci. Ec. Norm. Sup.* III **17**, 21–86, English translation in *The random character of stock market prices* (1964), (ed. P. H. Cootner), pp. 17–78. Cambridge Mass.: MIT Press.

Barnsley, M. F. & Elton, J. H. 1988 A new class of Markov processes for image encoding. *Advances appl. Prob.* **20**, 14–32.

Bick, A. 1987 On the consistency of the Black–Scholes model with a general equilibrium framework. *J. Financial Quant. Anal.* **22**, 259–275.

Black, F. 1986 Noise. *J. Finance* **XLI**, 529–543.

Black, F. & Scholes, M. 1973 The pricing of options and corporate liabilities. *J. Political Econ.* **81**, 637–659.

Brandt, A. 1986 The stochastic equation $Y_{n+1} = A_n Y_n + B_n$ with stationary coefficients. *Advances appl. Prob.* **18**, 211–220.

Brennan, M. J. & Schwartz, E. S. 1989 Portfolio insurance and financial market equilibrium. *J. Business* **62**, 455–476.

Brock, W. A. 1993 Pathways to randomness: emergent nonlinearity and chaos in economics and finance. *Estudios Economicos* **8**, 3–55.

Day, R. H. & Huang, W. 1990 Bulls, bears and market sheep. *J. Econ. Behavior Organization* **14**, 299–329.

De Long, J. B., Shleifer, A., Summers, L. H. & Waldmann, R. J. 1990*a* Noise trader risk in financial markets. *J. Political Econ.* **98**, 703–738.

De Long, J. B., Shleifer, A., Summers, L. H. & Waldmann, R. J. 1990*b* Positive feedback investment strategies and destabilizing rational speculation. *J. Finance* **44**, 793–805.

Diaconis, P. & Shahshahani, M. 1986 Products of random matrices and computer image generation. In *Random matrices and their applications* (ed. J. E. Cohen, H. Kesten & C. M. Newman), *Contemporary Mathematics 50*, pp. 173–182. American Mathematical Society.

Durlauf, S. 1993 Non-ergodic economic growth. *Rev. Econ. Stud.* **60**, 349–366.

Föllmer, H. 1974 Random economies with many interacting agents. *J. math. Econ.* **1**, 51–62.

Föllmer, H. & Schweizer, M. 1993 A microeconomic approach to diffusion models for stock prices. *Math. Finance* **3**, 1–23.

Frankel, J. A. & Froot, K. 1986 The dollar as an irrational speculative bubble: a tale of fundamentalists and Chartists. *Marcus Wallenberg Papers Int. Finance* **1**, 27–55.

Frey, R. & Stremme, A. 1993 Portfolio insurance and volatility. Department of Economics, University of Bonn (Discussion paper B-256).

Johnson, N. L. & Kotz, S. 1970 *Distributions in statistics, continuous univariate distributions*. Boston: Houghton Mifflin.

Kirman, A. 1993 On ants, rationality and recruitment. *Q. J. Econ.* **CVIII**, 137–156.

Kreps, D. 1982 Multiperiod securities and the efficient allocation of risk: a comment on the Black–Scholes option pricing model. In *The economics of uncertainty and information* (ed. J. McCall), pp. 203–232. University of Chicago Press.

Lebowitz, J. L., Maes, C. & Speer, E. R. 1990 Statistical mechanics of probabilistic cellular automata. *J. statist. Phys.* **59**, 117–170.

Orléan, A. & Robin, J.-M. 1991 Variability of opinions and speculative dynamics on the market of a storable good. CREA, Ecole Polytechnique, Paris (preprint).

Samuelson, P. A. 1964 Rational theory of warrant pricing. In *The random character of stock market prices* (ed. P. H. Cootner), pp. 506–525. Cambridge, Mass.: MIT Press.

Slominski, L. 1989 Stability of strong solutions of stochastic differential equations. *Stochastic Processes Applications* **31**, 173–202.

Smith, V. L., Suchanek, G. L. & Williams, A. W. 1988 Bubbles, crashes, and endogenous expectations in experimental spot asset markets. *Econometrica* **56**, 1119–1151.

Summers, L. 1986 Does the stock market rationally reflect fundamental values. *J. Finance* **XLI**, 591–601.

Vassershtein, L. N. 1969 Markov processes over denumerable products of spaces describing large systems of automata. *Problemy Peredaci Informacii* **5**, 64–72.

Discussion

W. Cheung (*Quorom Capital Management Ltd, London, U.K*). Recent work suggests that the volatility of equity markets are higher than that which would arise from pure fundamentals. The work of Robert Schiller, for example, on the US equity market compares actual volatility against the volatility of long-term fundamental values based on a dividend discount model. Can this volatility excess be explained by your models; for example, the Ornstein–Uhlenbeck process for information traders which reverts around a 'perceived fundamental' price process?

M. A. H. Dempster (*University of Essex, U.K.*). There is a class of stochastic specifications of security price processes, characterized by jumps and heavy tailed state distributions, different from those treated. They were first introduced by

Mandelbrot (1963) and may be interpreted as (geometric) brownian motion evaluated at random time changes. Mandelbrot & Taylor (1967) utilize a (paretian) time change from the stable class, while more recently Madan & Seneta (1994) propose a time change process with independent gamma increments. In both cases, a continuous time jump process results which in the second case is a semi-martingale (i.e. jumping diffusion). Vanilla option pricing results are available when these price processes are assumed to apply to the underlying security (see Rachev & Samorodnitsky 1991; Madan & Milne 1990). Mandelbrot & Taylor (1967) give an explanation of their compound price process which can be interpreted as real-time transactions bunching of virtual Black–Scholes trading. It would be interesting to relate this effect to the more general explanations in terms of agents' behaviour.

H. FÖLLMER. Stochastic security price processes are generated by the interaction of economic agents and that this interaction should be seen as an additional source of randomness, beyond any randomness in the fundamentals. In passing to the limit of continuous time we use the standard form of the invariance principle. It would indeed be interesting to consider various modifications of this step, e.g. to allow for random time changes.

Additional references

Madan, D. N. & Milne, F. 1990 Option pricing with V. G. martingale components. Department of Economics, University of Maryland (Working paper).

Madan, D. B. & Seneta, E. 1994 The V.G. model for share market returns. *J. Business*. (In the press.)

Mandelbrot, B. B. 1963 New methods in statistical economics. *J. Political Econ.* **71**, 421–440.

Mandelbrot, B. B. & Taylor, H. M. 1967 On the distribution of stock differences. *Operations Res.* **15**, 1057–1062.

Rachev, S. T. & Samorodnitsky, G. 1991 Option pricing formulae for speculative processes modelled by subordinated stochastic processes. Department of Mathematics, Cornell University (Tech. report 172).

4

A note on super-replicating strategies

M.H.A. Davis and J.M.C. Clark

The standard Black–Scholes option pricing methodology fails in the presence of transaction costs because portfolios that exactly replicate the option pay-off no longer exist. Several alternative approaches have been proposed; our purpose is to examine one of them which is based on the idea of 'super-replicating' portfolios. It is argued that this approach does not lead to a viable theory of option pricing in continuous time.

1. Introduction

The Black–Scholes theory of option pricing relies on the existence of replicating portfolios: given a certain initial endowment, an investor can form a dynamic portfolio in a risky stock and a risk-free bank whose value at some later time T exactly coincides with that of, for example, a European call option with exercise time T and some given strike price. An arbitrage opportunity exists for one party or the other if any other price than this initial endowment – the Black–Scholes price – is paid for the option at time 0. These matters are discussed in many textbooks, for example Cox & Rubinstein (1985) or Merton (1992). The argument fails when there are transaction costs or other forms of market friction, for then, replicating portfolios generally no longer exist. A variety of alternative approaches has been proposed to deal with this, the first important work being due to Leland (1985) who examined the effects of periodic rebalancing of a hedging portfolio. The purpose of the present paper is to take a critical look at another approach based on the idea of super-replicating strategies (SRS). An SRS for an option is a trading strategy such that the value of the corresponding portfolio is at least as great as that of the option at the exercise time. This idea is in fact used in the standard Black–Scholes framework in connection with pricing American options; see Karatzas (1989). In the discrete-time binomial model, perfect replication is still possible even in the face of transaction costs, but Bensaid *et al.* (1992) uncovered the intriguing fact that super-replication may be cheaper, providing therefore a tighter bound on feasible option prices.

We believe, however, that super-replicating strategies do not form the basis for a viable theory of option pricing in continuous time. For a call option, there is always one trivial SRS, namely to buy and hold one share, the corresponding initial endowment being of course the share price at time 0 plus any transaction costs on the purchase. We conjecture that this is the cheapest SRS. Although no proof of this is as yet available, there is strong evidence in favour of it. If true, option pricing based on SRS should be discarded in favour of alternative approaches based on utility maximization (see Hodges & Neuberger 1989 or Davis *et al.* 1993), which 'soften' the penalty for under-replication.

Mathematical Models in Finance. Edited by S.D. Howison, F.P. Kelly and P. Wilmott.
Published in 1995 by Chapman & Hall, London. ISBN 0 412 63070 2

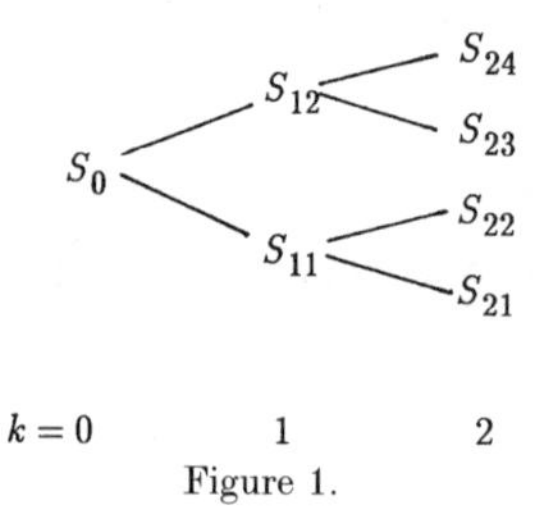

Figure 1.

The paper proceeds as follows. In §2 the discrete-time binomial model is described and the existence of replicating and possibly cheaper super-replicating strategies demonstrated. The continuous-time model including transaction costs is formulated in §3, following Davis & Norman (1990) and Davis *et al.* (1993). In the final section (§4) our conjecture is stated formally and the evidence for it presented.

2. Discrete-time binomial models

Option pricing based on a binomial stock price model was introduced by Cox *et al.* (1975) and is discussed in §10.2 of Merton (1992). Let us consider a 2-period model with time instants $k = 0$, 1, 2 and a single stock whose price evolves as shown in figure 1.

The initial price is $S(0) = S_0$ and this moves to $S(1) = S_{11}$ or S_{12} at time $k = 1$, etc. It is not necessary to specify the probabilities of these transitions: the only requirement is that every path in the tree must have strictly positive probability. By convention, $S_{11} < S_{12}$, $S_{21} < S_{22}$, $S_{23} < S_{24}$. There is also a riskless asset, the 'bank', which yields a return R per dollar over one period. To rule out arbitrage we clearly must have $S_{11} < RS_0 < S_{12}$, $S_{23} < RS_{12} < S_{24,}$, $S_{21} < RS_{11} < S_{22}$ as otherwise riskless profits can be made by borrowing from bank to invest in stock or vice versa.

Suppose that a contingent claim has value $h(S(2))$ at time $k = 2$. It is easy to see how to form a replicating portfolio for this. Let N_{11}, B_{11} $[N_{12}, B_{12}]$ denote the number of stock units and the amount invested in bank at time 1 when the stock price is $S(1) = S_{11}[= S_{12}]$. Then for perfect replication we must have

$$\left.\begin{aligned} RB_{11}+N_{11}S_{21} = h(S_{21}), \quad RB_{12}+N_{12}S_{23} = h(S_{23}), \\ RB_{11}+N_{11}S_{22} = h(S_{22}), \quad RB_{12}+N_{12}S_{24} = h(S_{24}). \end{aligned}\right\} \tag{2.1}$$

These equations are easily solved to give

$$\left.\begin{aligned} N_{11} &= [h(S_{21})-h(S_{22})]/(S_{21}-S_{22}), \\ N_{12} &= [h(S_{23})-h(S_{24})]/(S_{23}-S_{24}), \\ B_{11} &= [S_{21}H(S_{22})-S_{22}H(S_{21})]/R(S_{21}-S_{22}), \\ B_{12} &= [S_{23}H(S_{24})-S_{24}H(S_{23})]/R(S_{23}-S_{24}). \end{aligned}\right\} \tag{2.2}$$

The value of the hedging portfolio at time 1 is then $F_{11} = B_{11}+N_{11}S_{11}$, if $S(1) = S_{11}$, and $F_{12} = B_{12}+N_{12}S_{12}+N_{12}S_{12}$, if $S(1) = S_{12}$. We can now perform a similar analysis to determine N_0 and B_0, the number of stock units and dollars held in bond at time 0 so that F_{11} and F_{12} are perfectly replicated at time 1. The price of the contingent claim h at time 0 is then $F_0 = B_0+N_0S_0$, the value of the replicating portfolio.

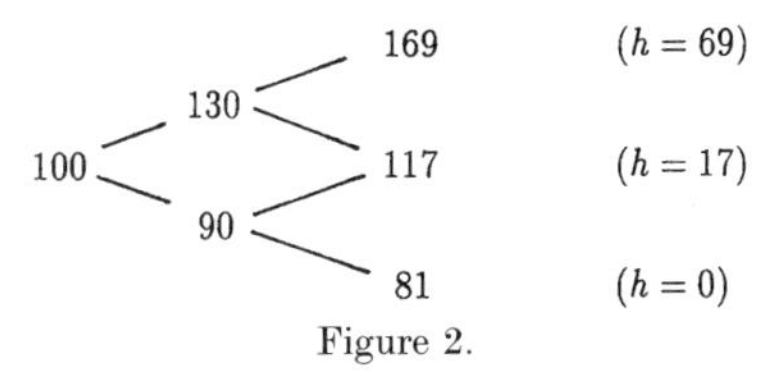

Figure 2.

As an example, take $R = 1$ and consider the values shown in figure 2. (At each stage the price moves from S to uS or dS where $u = 1.3$, $d = 0.9$.)

The contingent claim h is an at-the-money call option, so that $h(S(2)) = [S/2) - 100]^+$. (Notation: $[X]^+ = \max\{X, 0\}$.) We find that the option value at time 0 is $F_0 = 10.69$ and the initial composition of the replicating portfolio is $N_0 = 0.644$, $B_0 = -53.71$. The option value F_0 can be expressed as the discounted expectation with respect to a risk-neutral probability distribution, but we do not need this result here.

Perhaps surprisingly, perfect replication is still possible when transaction costs are introduced. This is covered in §14.2 of Merton (1992) and in Boyle & Vorst (1992) and Edirisinghe *et al.* (1993). Let us maintain the 2-period model of figure 1, but assume that a fraction λ of the amount transacted is paid on all movements in and out of stock and that settlement at time $k = 2$ is in cash. The latter is just a convention and the argument below works *mutatis mutandis* for other forms of settlement, for example delivery of one stock unit in return for a cash payment of K for an in-the-money call option with strike price K. Let N_{ij}^{λ}, B_{ij}^{λ} denote the number of units of stock and the dollars held in bond for the new model, so that $N_{ij} = N_{ij}^0$ etc. where N_{ij} is the quantity appearing in the original transaction-cost-free model. The requirements (2.1) for exact replication at time $k = 2$ now become

$$RB_{11}^{\lambda} + (1-\lambda) N_{11}^{\lambda} S_{21} = h(S_{21}), \quad RB_{12} + (1-\lambda) N_{12}^{\lambda} S_{23} = h(S_{23}),$$
$$RB_{12}^{\lambda} + (1-\lambda) N_{11}^{\lambda} S_{22} = h(S_{22}), \quad RB_{12} + (1-\lambda) N_{12}^{\lambda} S_{24} = h(S_{24}),$$

as the portfolio must be cashed out at time 2, giving

$$N_{11}^{\lambda} = (1/(1-\lambda)) N_{11}, \quad N_{12}^{\lambda} = (1/(1-\lambda)) N_{12},$$
$$B_{11}^{\lambda} = B_{11}, \quad B_{12}^{\lambda} = B_{12}.$$

Because $S_{11} < S_0 < S_{12}$ it is reasonable and, as it turns out, correct to suppose that $N_{11}^{\lambda} < N_0^{\lambda} < N_{12}^{\lambda}$. Thus the portfolio value required before rebalancing at time $k = 1$ is, when $S(1) = S_{11}$,

$$\begin{aligned} F_{11}^{\lambda} &= S_{11} N_{11}^{\lambda} + B_{11}^{\lambda} + \lambda (N_0^{\lambda} - N_{11}^{\lambda}) S_{11} \\ &= (1-\lambda) S_{11} N_{11}^{\lambda} + B_{11}^{\lambda} + \lambda N_0^{\lambda} S_{11} \\ &= F_{11} + \lambda N_{11}^{\lambda} S_{11}. \end{aligned}$$

When $S(1) = S_{12}$ the corresponding amount is

$$\begin{aligned} F_{12}^{\lambda} &= S_{12} N_{12}^{\lambda} + B_{12}^{\lambda} + \lambda (N_{12}^{\lambda} - N_0^{\lambda}) S_{12} \\ &= F_{12} + \lambda (2N_{12}^{\lambda} - N_0^{\lambda}) S_{12}. \end{aligned}$$

To provide these values, we must have

$$RB_0^{\lambda} + N_0^{\lambda} S_{11} = F_{11} + \lambda N_0^{\lambda} S_{11},$$
$$RB_0^{\lambda} + N_0^{\lambda} S_{12} = F_{12} + \lambda (2N_{12}^{\lambda} - N_0^{\lambda}) S_{12},$$

giving

$$N_0^\lambda = (F_{12} - F_{11} + 2\lambda N_{12}^\lambda S_{12})/[(1+\lambda)S_{12} - (1-\lambda)S_{11}],$$

$$B_0^\lambda = [(1+\lambda)F_{11}S_{12} - (1-\lambda)F_{12} - 2\lambda(1-\lambda)N_{12}^\lambda S_{12}S_{11}]/[R(1+\lambda)S_{12} - (1-\lambda)S_{11})].$$

The cash required to set up this portfolio is then

$$F_0^\lambda = B_0^\lambda + (1+\lambda)N_0^\lambda S_0.$$

Thus F_0^λ is an upper bound for the price a buyer would be prepared to pay for the contingent claim, since at any higher price he or she is certainly providing the writer with a riskless profit.

Let us take (just for illustrative purposes) $\lambda = 0.2$ in the example of figure 2. Then we find that $N_0^\lambda = 1.080$ and $B_0^\lambda = -30.68$. Thus the initial value of the replicating portfolio is 77.32 ignoring the set-up cost, or $F_0^\lambda = 98.92$ including the set-up cost. However, as pointed out by Bensaid *et al.* (1992), there are strategies that are clearly better than this. For example, suppose we take $N_0 = N_0^\lambda = 1.08$, $B_0 = -77$ and do no trading at time $k = 1$. Then the cashed-out values of the portfolio at $k = 2$ are 69.02 and 24.01, when $S(2) = 169$, 117 respectively, whereas the 'book' value of the portfolio is 10.48 when $S(2) = 81$. Because, in the latter case, nothing has to be delivered, there is no reason to cash out the portfolio, and the position is therefore more than covered at all values of $S(2)$, at an initial cost of 52.6 (or 31 without the set-up costs), far less than the cost of the replicating portfolio.

One of the main uses of the binomial formulation of option pricing theory is as an approximation to the continuous-time case in which, as outlined in §3 below, the stock price is modelled as a geometric brownian motion with volatility (or standard deviation) parameter σ. This is equivalent to approximating the logarithm of the stock price by a discrete random walk. We take an n-period model in which $S(k+1)$ is equal to $uS(k)$ or $dS(k)$ where $d < 1 < u$. If in the continuous time model the trading interval is $[0, T]$ and the riskless interest rate r then the appropriate scaling is obtained by taking $h = T/n$, $u = e^{\sigma\sqrt{h}}$, $d = e^{-\sigma\sqrt{h}}$ and $R = e^{rh}$. This approximation is described very clearly in Cox & Rubinstein (1985). In an ingenious analysis, Boyle & Vorst (1992, Theorem 3) show that if one applies this scaling to the binomial model with transaction costs, then for large n and small λ the initial value of the replicating portfolio for a call option is approximately equal to the Black–Scholes value but with a modified variance given by

$$\sigma_n^2 = \sigma^2[1 + (2\lambda/\sigma)\sqrt{(n/T)}].$$

Recall that the 'delta' (i.e. the number of stock units) in the Black–Scholes replicating portfolio is $\Delta(T, S) = N(d_1)$ where N is the cumulative normal distribution function and

$$d_1 = [\ln(S/K) + (r + \tfrac{1}{2}\sigma^2)(T-t)]/\sigma\sqrt{(T-t)}.$$

Thus $\Delta(T, S) \to 1$ as $\sigma \to \infty$ and Boyle & Vorst's results show that as the number of periods increases the initial position in the stock converges to one while the position in bond converges to zero. Hence the replicating portfolio is for most periods close to a buy-and-hold policy of simply purchasing a unit of stock at time 0. Such a policy clearly covers the option position at time T.

Based on the above considerations, Bensaid *et al.* (1992) analyse hedging problems with transaction costs, by considering super-replicating strategies (SRS). Suppose $h(S_T)$ is the value of a European contingent claim at exercise time T, S_T being the stock price at that time. An SRS is a dynamic portfolio whose value X_T at time T satisfies $X_T \geqslant h(S_T)$ with probability one. We have already seen that – at first sight paradoxically – the cheapest SRS may be cheaper than the initial cost of the replicating portfolio if transactions costs are present. Thus the SRS concept gives tighter bounds on option values. Bensaid *et al.* (1992) give a constructive procedure for calculating the cheapest SRS in the binomial model. They do not however consider asymptotics, i.e. what happens when the model is rescaled and the number of periods increases to infinity as described above. Rather than do this, we will, in the following sections, examine SRS directly in a continuous-time setting.

3. A continuous-time model with transaction costs

Our continuous-time model, which is similar to the one used by Davis & Norman (1990), is a slight variation of the standard geometric brownian motion model. Let (w_t) be a brownian motion with natural filtration $(\mathscr{F}_t)$, defined on some probability space $(\Omega, \mathscr{F}, P)$. The market contains a single risky asset ('stock') whose price (S_t) satisfies

$$\mathrm{d}S_t = \alpha S_t\,\mathrm{d}t + \sigma S_t\,\mathrm{d}w_t, \quad S_0 > 0.$$

Here α, $\sigma > 0$ are given constants. There is also a riskless asset ('bank') paying a constant interest rate r. Let (y_t) and (z_t) be the processes of dollar holdings in the stock and bank respectively. A trading strategy is a pair (L_t, U_t) of $(\mathscr{F}_t)$-adapted, right-continuous increasing processes, and the evolution of portfolio holdings (y_t, z_t) in response to such a trading strategy is defined by the following equations:

$$\begin{aligned} \mathrm{d}y_t &= \alpha y_t\,\mathrm{d}t + \sigma y_t\,\mathrm{d}w_t + \mathrm{d}L_t - \mathrm{d}U_t, & y_0 &= y, \\ \mathrm{d}z_t &= r z_t\,\mathrm{d}t - (1+\lambda)\,\mathrm{d}L_t + (1-\lambda)\,\mathrm{d}U_t, & z_0 &= z, \end{aligned}$$

where (y, z) is the initial portfolio composition and $\lambda \geqslant 0$ is the transaction cost parameter. Note that purchase of $\$\mathrm{d}L$ of stock requires a payment of $\$(1+\lambda)\,\mathrm{d}L$ from bank, while sale of $\$\mathrm{d}U$ of stock realizes only $\$(1-\lambda)\,\mathrm{d}U$ in cash. Thus λ is the fractional transaction cost and all transaction costs are paid from bank. (Note that in the model of Davis *et al.* (1993), (y_t) represents the number of stock units rather than their dollar value.)

To simplify the notation in the following discussion we will, without essential loss of generality, take $r = \alpha = 0$ and $\sigma = 1$. Taking $r = 0$ and $\sigma = 1$ just amounts to a specific choice of numeraire and monetary units, whereas $\alpha = 0$ can always be achieved by replacing P by an equivalent (risk neutral) probability measure. Most of the following argument is concerned with 'almost sure' properties, which are invariant under mutually absolutely continuous substitution of measures. To summarize, our market model is

Stock: $$\mathrm{d}S_t = S_t\,\mathrm{d}w_t, \tag{3.1}$$

\$ holdings in stock: $$\mathrm{d}y_t = y_t\,\mathrm{d}w_t + \mathrm{d}L_t - \mathrm{d}U_t, \quad y_0 = y, \tag{3.2}$$

\$ holdings in bank: $$\mathrm{d}z_t = -(1+\lambda)\,\mathrm{d}L_t + (1-\lambda)\,\mathrm{d}U_t = \mathrm{d}A_t - \lambda \mathrm{d}\breve{A}_t, \quad z_0 = z. \tag{3.3}$$

In (3.3), $A_t := U_t - L_t$ is the net sale of stock and $\breve{A}_t := U_t + L_t$. It is clearly pointless to buy and sell stock at the same time, so we may and shall restrict ourselves to trading strategies such that $(\breve{A}_t)$ is the total variation of the bounded variation transaction process (A_t).

We now describe, following Davis & Norman (1990) or Davis *et al.* (1993), the sort of trading strategies that are optimal from the point of view of maximizing utilities or other functions of the vector of portfolio holdings. Thus, suppose $T > 0$ is a fixed time and we wish to maximize $\mathbb{E}\Phi(s_T, y_T, z_T)$ where Φ is some given function and $\mathbb{E}$ denotes expectation with respect to the probability measure P. For example, taking $\Phi(S,y,z) = (1-\lambda)y+z$ corresponds to maximizing the expected cash value of the portfolio at time T, while $\Phi(S,y,z) = (1-\lambda)y+z-[S-K]^+$ corresponds to maximizing this value if a call option with strike price K has to be settled first. Suppose first that (L_t, U_t) are restricted to be absolutely continuous with derivative bounded by κ, i.e. $dL_t = l_t\,dt, dU_t = u_t\,dt$ and $0 \leqslant l_t, u_t \leqslant \kappa$. Then we have a 'standard' stochastic control problem, and formal dynamic programming arguments (see Fleming & Rishel, 1975) indicate that the value function

$$V(t,S,y,z) = \sup_{(L,U)} \mathbb{E}_{t,(S,y,z)}\,\Phi(S_T, y_T, z_T) \tag{3.4}$$

should satisfy the Bellman equation

$$\frac{\partial V}{\partial t}+\tfrac{1}{2}S^2\frac{\partial^2 V}{\partial S^2}+\tfrac{1}{2}y^2\frac{\partial^2 V}{\partial y^2}+Sy\frac{\partial V}{\partial S\,\partial y}$$
$$+\max_{l,\,u\in[0,\kappa]}\left\{\left[\frac{\partial V}{\partial y}-(1+\lambda)\frac{\partial V}{\partial z}\right]l+\left[-\frac{\partial V}{\partial y}+(1-\lambda)\frac{\partial V}{\partial z}\right]u\right\}=0 \tag{3.5}$$

with boundary condition

$$V(T,S,y,z) = \Phi(S,y,z). \tag{3.6}$$

The key point in (3.5) is that the maxima are achieved at the extreme values $\{0,\kappa\}$, the values taken depending on the signs of the terms multiplying l, u in (3.5). Specifically the (t, S, y, z) space splits into three regions:

$\mathscr{B}$ (buy at maximum rate, $l=\kappa, u=0$): $\partial V/\partial y \geqslant (1+\lambda)\partial V/\partial z$,

$\mathscr{S}$ (sell at maximum rate, $l=0, u=\kappa$): $\partial V/\partial y \leqslant (1-\lambda)\partial V/\partial z$,

$\mathscr{NT}$ (no transactions, $l=u=0$): $(1-\lambda)\partial V/\partial z < \partial V/\partial y < (1+\lambda)\partial V/\partial z$.

As $\kappa\to\infty$ this split is maintained, but the transactions in $\mathscr{B}$ and $\mathscr{S}$ take place at 'infinite speed', corresponding to an instantaneous transaction to the boundary of the no-transaction region, implying that the above inequalities in $\mathscr{B}$, $\mathscr{S}$ hold with equality throughout these regions. In the limit (3.5) becomes the variational inequality

$$\max\left\{(1-\lambda)\frac{\partial V}{\partial z}-\frac{\partial V}{\partial y},-(1+\lambda)\frac{\partial V}{\partial z}+\frac{\partial V}{\partial y},\frac{\partial V}{\partial t}+\tfrac{1}{2}S^2\frac{\partial^2 V}{\partial S^2}+\tfrac{1}{2}y^2\frac{\partial^2 V}{\partial y^2}+yS\frac{\partial V}{\partial S\,\partial y}\right\}=0, \tag{3.7}$$

with the same boundary condition (3.6) as before, to be solved in a certain 'solvency region' of (t, S, y, z) space; see Davis *et al.* (1993). This equation is 'degenerate' in that the second order term is not uniformly elliptic (there is no 'noise' in the z-component), and it therefore cannot *a priori* be expected to have a classical ($C^{1,2}$) solution. However, under appropriate conditions on Φ, the variational inequality has a unique solution in the viscosity sense (see Fleming & Soner (1992) for this) which coincides with V defined by (3.4). After a possible initial jump to the boundary of the no-transaction region $\mathscr{NT}$, the optimal transaction processes (L_t, U_t) are the local times of the portfolio process at the boundaries $\partial\mathscr{B}, \partial\mathscr{S}$ respectively. These are the minimal increasing processes needed to keep the portfolio vector within $\mathcal{NT}$; they are continuous but singular with respect to Lebesgue measure and increase only on the boundaries $\partial\mathcal{B}, \partial\mathcal{S}$. The optimal portfolio is in the interior of $\mathcal{NT}$, where $\mathrm{d}L = \mathrm{d}U = 0$, i.e. no transactions take place, almost all of the time.

4. Super-replicating strategies

Let us now consider a European call option on the stock (S_t) with exercise time T and strike price K. For simplicity of exposition we will assume that transactions are free at times $0, T$. Thus the value of the option at exercise is $[S(T)-K]^+$ and an initial cash amount z can be invested to form a portfolio whose initial composition is $(z-z', z')$ for any z'; after this, transactions incur a proportional charge λ as described in §3.

Definition 4.1. **A** *A super-replicating strategy* (SRS) *is an initial cash endowment z and a trading strategy (L, U) such that*

$$y_T + z_T \geqslant [S_t - K]^+ \quad a.s.,$$

where (y_t, z_t) is the solution of (3.2), (3.3) *with $y_0 = z - z', z_0 = z'$ for some $z' \in \mathbb{R}$.*

There is one trivial SRS, namely to buy and hold one share. This corresponds to taking $z = S_0$, $z' = 0$, $L_t \equiv U_t \equiv 0$ so that $y_T + z_T = S_T \geqslant [S_T - K]^+$. Based on the behaviour of the discrete-time models discussed in §§1 and 2, the fundamental contention of this paper is

Conjecture 4.2. *The trivial buy-and-hold strategy is the cheapest super-replicating strategy when the transaction cost parameter λ is strictly positive.*

A proof of this statement has eluded us so far, but we have found no counter-example and a considerable amount of evidence in favour of it, which is adduced in the remainder of this section.

Because the buy-and-hold strategy has cost S_0, the conjecture amounts to saying that no strategy with $z < S_0$ is super-replicating. In view of the free transaction at $t = 0$ we can take as initial conditions in (3.2), (3.3) $y_{0-} = S_0 - \epsilon, z_{0-} = 0$ for some $\epsilon > 0$ and the conjecture is established if $P[y_T + z_T < [S_T - K]^+] > 0$ for every trading strategy (L_t, U_t) with this initial condition.

Example 4.3. A strategy that 'nearly' works is to take $L_t = \epsilon$ (i.e. $\Delta L_0 = \epsilon$), $U_t = 0$, i.e. we borrow ϵ from bank and buy a share; then $y_t = S_t$ and $z_t = -\epsilon$, and, assuming $\epsilon < K$, $y_T + z_T < [S_T - K]^+$ on the set $S_T < \epsilon$ which of course has positive probability. Insisting on super-replication with probability one is a very stringent condition.

Using the elementary identity $S \wedge K = S - [S-K]^+$ where $S \wedge K = \min\{S, K\}$ we have $y_T + z_T < [S_T - K]^+$ if and only if $\zeta_T > S_T \wedge K$, where ζ_T is the deficit process

$$\zeta_T = S_t - y_t - z_t,$$

the difference between the value of one share and the 'raw' value of the portfolio. Conjecture 4.2 is therefore established if, with $z_{0-} = 0$, $y_{0-} = S - \epsilon$,

$$P[\zeta_T > S_T \wedge K] > 0$$

or, equivalently,

$$\operatorname*{ess\,sup}_{\omega \in \Omega} (\zeta_T(\omega) - S_T(\omega) \wedge K) > 0. \tag{4.1}$$

This statement can be recast in terms of expectations in the following way. For a bounded adapted process, ψ_t, let Λ_T be the Girsanov density

$$\Lambda_T(\psi) = \exp\left(\int_0^T \psi_s \,\mathrm{d}w_s - \frac{1}{2}\int_0^T \psi_s^2 \,\mathrm{d}s\right).$$

The following lemma is easily established.

Lemma 4.4. *For any integrable $\mathscr{F}_T$-measurable random variable X we have*

$$\operatorname*{ess\,sup}_{\omega\in\Omega} X(\omega) = \sup_{\psi \text{ bounded}} E[\Lambda_T(\psi)X].$$

We now calculate $E[L(\psi)X]$ when X is given by the expression at (4.1).

Theorem 4.5. *Consider an arbitrary continuous trading strategy (L, U) with initial endowment $z_{0-} = 0$, $y_{0-} = S - \epsilon$. Then for any bounded measurable process (ψ_t),*

$$\mathbb{E}\{\Lambda_T(\zeta_T - S_T \wedge K)\} = \epsilon - g(0, S_0) + \tfrac{1}{2}\mathbb{E}\int_0^T \Lambda\psi(\phi - g'S)\,\mathrm{d}t + \lambda\,\mathbb{E}\int_0^T \Lambda\,\mathrm{d}\breve{A}_t. \tag{4.2}$$

In this equation $\Lambda_t = \Lambda_t(\psi)$, $\phi_t = S_t - y_t$ and $g' = (\partial g/\partial S)(t, S_t)$ where g is the unique solution of the partial differential equation

$$\partial g/\partial t + \tfrac{1}{2}S^2\,\partial^2 g/\partial S^2 = 0, \quad S > 0, \quad t \in [0, T], \quad g(T, S) = S \wedge K.$$

Proof. By standard arguments, $g(t, S) = \mathbb{E}_{t,S}[S_T \wedge K]$. Thus $g(t, s) = S - p(t, S)$, where $p(t, S)$ is the Black–Scholes price of the call option. From (3.1)–(3.3), $\phi_t = S_t - y_t$ satisfies

$$\mathrm{d}\phi_t = \phi_t\,\mathrm{d}w_t + \mathrm{d}A_t, \quad \phi_{0-} = \epsilon, \tag{4.3}$$

whereas $z_t = A_t - \lambda\breve{A}_t$, so that $\zeta_t = S_t - y_t - z_t = \phi_t - z_t$ satisfies

$$\mathrm{d}\zeta_t = \phi_t\,\mathrm{d}w_t + \lambda\,\mathrm{d}\breve{A}_t, \quad \zeta_{0-} = \epsilon. \tag{4.4}$$

Also the Girsanov density $\Lambda_t = \Lambda_t(\psi)$ satisfies

$$\mathrm{d}\Lambda_t = Lv\,\mathrm{d}w, \quad \Lambda_0 = 1, \tag{4.5}$$

whereas from the Itô formula and (4.3) we have

$$\mathrm{d}g(t, S_t) = g'S\,\mathrm{d}w. \tag{4.6}$$

As A_t is assumed to be continuous we have, by applying the Itô formula to (4.4)–(4.6),

$$\mathrm{d}(\Lambda_t\zeta_t) = \lambda(\phi\,\mathrm{d}w + \lambda\,\mathrm{d}\breve{A}) + \zeta\Lambda\psi\,\mathrm{d}w + \tfrac{1}{2}\Lambda\psi\phi\,\mathrm{d}t,$$
$$\mathrm{d}(\Lambda_t g(t, S_t)) = \Lambda g'S\,\mathrm{d}w + gL\psi\,\mathrm{d}w + \tfrac{1}{2}L\psi g'S\,\mathrm{d}t,$$

and hence

$$\Lambda_T(\zeta_T - S_T \wedge K) = \Lambda_T(\zeta_T - g(T, S_T))$$
$$= \epsilon - g(0, S_0) + \frac{1}{2}\int_0^T \Lambda\psi(\phi - g'S)\,\mathrm{d}t + \int_0^T L(\zeta\psi\phi - g'S - gv)\,\mathrm{d}w$$

The last term is a martingale when ψ is bounded, and (4.2) follows. □

Define $\alpha_t := \phi_t - g'(t, S_t) S_t$. Then Theorem 4.4 states that super-replication fails if and only if there is some process (ψ_t) such that

$$\mathbb{E}\int_0^T \Lambda_t \psi_t \alpha_t \, dt + 2\lambda \mathbb{E}\int_0^T \Lambda \, d\check{A}_t > 2(g(0, S_0) - \epsilon). \tag{4.7}$$

The process (α_t) has an interesting interpretation. Since $g = S - p$ and $\phi = S - y$,

$$\alpha_t = p'(t, S_t) S_t - y_t.$$

Thus α_t is the difference between the dollar value of stock holdings in the Black–Scholes replicating portfolio and in the portfolio corresponding to (L, U). In fact, it is impossible to 'track' the Black–Scholes portfolio exactly with bounded variation strategies, as we now show.

Proposition 4.6. *There is no trading strategy* (L, U) *such that* $\alpha_t \equiv 0$.

Proof. We find, using the Itô formula and (4.3), that

$$d(g'S_t) = -\tfrac{1}{2}S^2 g'' \, dt + (g'S + g''S^2) \, dw, \tag{4.8}$$

whereas

$$d\phi_t = dA_t + \phi \, dw. \tag{4.9}$$

If $\phi_t \equiv g'S_t$ then as A has bounded variation it must be the case that $dA_t = -(\tfrac{1}{2}S^2 g'') \, dt$ and $\phi = (g'S + g''S^2)$. But since $\phi = g'S$ the latter equality holds only if $g''(t, S_t) = 0$, whereas from the Black–Scholes formula we know that $g''(t, S) \neq 0$ for any (t, S). □

Just to check (4.7), note that if $\lambda = 0$ and we drop the requirement that A has bounded variation then a strategy A giving $\alpha_t = 0$ can be obtained simply by equating the right-hand sides of (4.8) and (4.9). For this strategy (4.2) implies that super-replication fails if and only if $g(0, S_0) - \epsilon < 0$ or, equivalently, $S_0 - \epsilon < p(0, S_0)$, i.e. if and only if the initial value of the portfolio is less than the Black–Scholes price.

On the positive side, there is one case that is easily disposed of.

Proposition 4.7. *Let* $A = (L, U)$ *be a trading strategy with initial portfolio as in Theorem* 4.4, *and define* α_t *by* (4.8). *Suppose that*

$$\int_0^T \frac{1}{\alpha_t^2} \, dt < \infty \quad a.s.$$

and that $\mathbb{E}\Lambda_T(\psi) = 1$ *when* $\psi_t = 1/\alpha_t$. *Then* (L, U) *is not a super-replicating strategy.*

Proof. Define $\psi_t = a/\alpha_t$ for some $a > 0$. Then $\mathbb{E}\Lambda_T(\psi) = 1$ under the stated conditions and

$$\mathbb{E}\int_0^T \Lambda_t \psi_t \alpha_t \, dt = \mathbb{E}\Lambda_T \int_0^T \psi_t \alpha_t \, dt = aT.$$

Thus inequality (4.7) is satisfied if a is sufficiently large. □

Corollary 4.8. *Super-replication fails if there is some* $\delta > 0$ *such that* $|\alpha_t| > \delta$ *for almost all* (t, ω).

Bearing in mind the description of optimal strategies in §3, the above propositions indicate that the only possible candidates for super-replicating strategies are those that track the Black–Scholes portfolio closely (i.e. keep $|\alpha_t|$ small) by the introduction of suitable reflecting barriers. On the other hand, if the sell and buy barriers are too close together then the transaction cost term in (4.7) will become large and super-replication will fail. To get some hints about the latter phenomenon, consider reflecting brownian motion $x_t = w_t + \xi_t^1 - \xi_t^2$ in an interval $[-b, b]$, where ξ_t^1, ξ_t^2 are the local times at $-b, b$ respectively. An easy argument shows that for $\delta > 0$,

$$\mathbb{E}_x \int_0^\infty \mathrm{e}^{-\delta t}\,(\mathrm{d}\xi_t^1 + \mathrm{d}\xi_t^2) = \frac{\cosh\sqrt{\delta}x}{\sqrt{\delta}\sinh\sqrt{\delta}b} =: v_\delta(x).$$

Now $\lim_{\delta\downarrow 0} \delta v_\delta(x) = 1/b$, showing by the 'final value theorem' of Laplace transforms that for small b,

$$\mathbb{E}_x(\xi_T^1 + \xi_T^2) \approx T/b.$$

Thus certainly the transaction cost term in (4.7) 'blows up' as tighter tracking of the Black–Scholes portfolio is attempted. It is a delicate question whether barriers can be devised that are wide enough apart to keep the second term in (4.7) small while being close enough together so that super-replication is not ruled out by Proposition 4.6. We think not.

References

Bensaid, B., Lesne, J.-P., Pagès, H. & Scheinkman, J. 1992 Derivative asset pricing with transaction costs. *Math. Finance* **2**, 63–86.

Boyle, P. P. & Vorst, T. 1992 Option replication in discrete time with transaction costs. *J. Finance* **47**, 272–293.

Cox, J. C., Ross, S. A. & Rubinstein, M. 1979 Option pricing: a simplified approach. *J. Financial Econ.* **7**, 229–263.

Cox, J. C. & Rubinstein, M. 1985 *Options markets*. Englewood Cliffs, New Jersey: Prentice-Hall.

Davis, M. H. A. & Norman, A. 1990 Portfolio selection with transaction costs. *Math. Operational Res.* **15**, 676–713.

Davis, M. H. A., Panas, V. G. & Zariphopoulou, T. 1993 European option pricing with transaction costs. *SIAM J. Control Optimization* **31**, 470–493.

Edirisinghe, C., Naik, V. & Uppal, R. 1993 Optimal replication of options with transaction costs and trading restrictions. *J. Financial Quant. Anal.* **28**, 117–138.

Fleming, W. H. & Rishel, R. W. 1975 *Deterministic and stochastic optimal control*. New York: Springer-Verlag.

Fleming, W. H. & Soner, H. M. 1992 *Controlled diffusion processes and viscosity solutions*. New York: Springer-Verlag.

Hodges, S. D. & Neuberger, A. 1989 Optimal replication of contingent claims under transaction costs. *Rev. Futures Markets* **8**, 222–239.

Karatzas, I. 1989 Optimization problems in the theory of continuous trading. *SIAM J. Control Optimization* **27**, 1221–1259.

Leland, H. E. 1985 Option pricing and replication with transaction costs. *J. Finance* **40**, 1283–1301.

Merton, R. C. 1992 *Continuous time finance* (revised edn). Cambridge, Mass.: Basil Blackwell.

5

Worldwide security market anomalies

W. T. ZIEMBA AND C. R. HENSEL

Systematic violations of security market efficiency occur in equity markets because of the timing and reaction to cash flows and other information, institutional constraints and policies, and investor behaviour. They lead to significantly different risk-adjusted returns to those expected. Taking these anomalies into account provides opportunity for superior investment performance.

Classifying anomalies as fundamental or seasonal differentiates between individual securities and market timing with indices. Seasonal anomalies include the January small-firm, turn-of-the-month, holiday, and day-of-the-week effects. Seasonality calendars combine the various effects to provide daily return forecasts.

Fundamental anomalies include price to earnings, price to book, market capitalization, dividend yield, earnings trends and surprises, and mean reversion effects. These variables add explanatory power to that from risk measures and yield factor models to separate the best from the worst performing individual securities.

Anomalies are controversial, difficult to measure and variable in time through investor sentiment and futures anticipation. Their study is interesting and challenging, and they are useful in various areas of portfolio management.

---

1. Introduction

One of the most basic and useful paradigms of modern financial economics is the efficient market hypothesis. In a pure form the hypothesis may be stated as: current security prices reflect all publicly available information. This, surely, is false. More appropriate for actual markets is that current security prices reflect all publicly available information to the point where the marginal profits from acting on the information do not exceed the marginal costs. In other words, current prices are within a transactions cost band. In a wide variety of equity, commodity and other financial markets, this is more or less true. However, rigorous and defensible tests of this version of the hypothesis are difficult to carry out and equally hard to interpret. First there is a messy problem of dealing with transactions costs. Moreover, as stressed by Roll (1977), such a test, if properly undertaken, should be a joint hypothesis concerning market efficiency and the validity of an equilibrium pricing model, such as the Sharpe–Lintner capital asset pricing model or Ross's arbitrage pricing model.

The British statistician, Maurice Kendall, postulated the efficient market hypothesis in 1953. Price changes were assumed to follow a random walk. Kendall (1953) found considerable support for the hypothesis by examining British stock

Mathematical Models in Finance. Edited by S.D. Howison, F.P. Kelly and P. Wilmott.
Published in 1995 by Chapman & Hall, London. ISBN 0 412 63070 2

prices as well as US commodity prices. Kendall's work ushered in the efficient market era which is well represented by the important early studies by Roberts (1959, 1967), Black *et al.* (1972), Cootner (1964), Fama (1965) and Mandelbrot (1963). The early literature was surveyed by Fama (1970).

Over the years market efficiency has evolved from studying weak, semi-strong and strong form efficiency (in which prices fully reflect price information, all publicly available information and all information respectively) to studies of the predictability of returns based on fundamental variables, event studies of the adjustment of prices to specific public announcements and tests for private information. The evidence is that future returns are predictable from past returns, dividend yields and term-structure variables. This apparent violation of weak form efficiency of the constant expected returns model is confounded by the joint hypothesis problem of whether there is rational variation over time in expected returns or systematic deviations from fundamental value. Various security market regularities or anomalies can then be seriously analysed, as surveyed in Dimson (1988), Thaler (1992) and Ziemba (1988, 1994*a*), and the review articles by Fama (1991), Blume & Siegel (1992), Hawawini & Keim (1994) and Ziemba (1994*b*).

Anomalies that yield higher short-term equity returns seem to occur for six basic reasons: (*a*) increased cash flows just prior to and during the anomalous period, a portion of which is invested in equities; (*b*) institutional constraints and policies such as pension funds investments made on the last day of the month; (*c*) behavioural considerations such as investor sentiment leading to excess purchase or sale of equities in related but different securities; (*d*) the timing of favourable and unfavourable information flows such as the delays in reporting bad news; (*e*) market maker supply–demand balances and bid–ask spread preferences; (*f*) the slowness of the market to react to new information such as earnings changes, favourable and unfavourable publicly available ranking changes. Whereas arguments may be made that increased returns occur because of increased risk, which is difficult or impossible to measure accurately, there is very strong evidence that most or all the gains in equity markets actually occur during a small subset of the year's trading days, namely the seasonally anomalous periods. Ritter & Chopra (1989) and Cadsby (1992) show, for example, that the only periods where higher risk, as measured by the capital asset pricing model's beta, is rewarded with higher equity returns is precisely at the anomalous periods such as the trading day before holidays, at the end of the week, at the turn of the month, in the first two weeks of January for small stocks, etc. Event studies are more straightforward and less controversial as they are able to provide more clear-cut evidence of the effect of the new information. Regarding strong form tests there is considerable evidence that corporate insiders have private information that is not fully reflected in current prices.

In the study of anomalies several key questions must be considered.

1. What are the anomalies, and do they really exist as opposed to their being simply random data perturbations discovered because many data sets have been investigated simultaneously? See Merton (1985), Lo & MacKinlay (1990), Black (1986, 1992) and Markowitz (1992) for discussion of this point.

2. Why do they occur?

3. How reliable are they?

4. How can you tell when the conditions that led to the existence of the anomaly have changed so that the anomaly no longer exists and prices are completely random around some expected value?

5. Is the anomaly in question anticipated or altered in the futures, options or other derivative security markets?

6. Can investors profit from such anomalies in a speculative sense?

7. Can investors use anomalies in an organized way in portfolio management to achieve superior risk-adjusted performance?

It is a large task to survey and discuss all the various anomalous areas, and our space is short. Hence we concentrate on one very important anomaly, namely the turn-of-the-month effect, and trace some of the results in worldwide markets. Along the way, evidence will be presented concerning questions 1–7. We refer the reader to the detailed survey papers mentioned above for specifics of further modelling, fundamental anomalies based on concepts such as price–earnings ratios, earnings surprise, etc., and other seasonal anomalies. The premier seasonal anomaly is the January small firm effect. It has been studied by many authors such as Banz (1981), Reinganum (1981), Blume & Stambaugh (1983), Keim (1983), Clark & Ziemba (1987), Jaffe, Keim & Westerfield (1989), Fama (1991), Prahl (1993) and Ziemba (1994*c*). Some of the turn-of-the-month discussion here will consider separately the unique month of January.

2. The turn-of-the-month effect in US equity markets

Investment advisors such as Fosback (1976) and Hirsh (1986) have argued that US stocks have substantial rises at the turn of the month. Ariel (1987) has documented the effect for small and large capitalization stocks for the 19 years from 1963 to 1981. His data consisted of the equal-weighted and value-weighted indices of all New York Stock Exchange (NYSE) stocks from the Center for Research on Security Prices (CRSP) tape. The turn of the month (TOM) is defined to be the last trading day of the previous month (-1) and the first four trading days of the new month ($+1$ to $+4$). Ariel's research showed there were very high returns during TOM. The rest of the market gains during 1963–1981 occurred in the second week of the month (days $+5$ to $+9$). The first half of the month (FH), namely trading days -1 to $+9$ or TOM plus the second week, had all the gains. The second half of the month (ROM, for the rest of the month), which is trading days $+10$ to -2, had negative returns. Hence, investment in the first half of the month provided more than all the year's stock market gains.

Lakonishok & Smidt (1988) investigated various seasonal anomalies using a 90 year data set on the Dow Jones Industrial Average (DJIA) from 1897 to 1986. The DJIA is a large capitalization price-weighted index of 30 major NYSE stocks. They found that the index rose 0.475% during the four-day period of -1 to $+3$ each month, whereas the average gain for a four-day period was 0.061%. The average gain per month over these 90 years was 0.349%. Hence, aside from these four days at the turn of the month, the DJIA had negative returns.

Ogden (1987, 1990) provides empirical support for the hypothesis about flow of funds into the stock market from cash flows and monetary action of agencies such as the Federal Reserve System. According to Moody's Manuals, 70% of the interest and principal payments on corporate debt (90% on municipal debt) are payable on the first or last days of the month. Furthermore, data in Standard & Poor's Stock Guide indicate that about 45% of dividends on common stock (65% on preferred stock) are payable on the first or last day of the month. The payable dates for interest, principal and dividend payments on corporate and municipal securities have been consistently on these dates throughout the 20th Century. Ogden's research provides empirical support for the hypothesis that stock prices tend to rise at the beginning of months that are preceded by months in which aggregate, economy-wide liquid profits are large. In particular, year-end bonuses, large Christmas sales (which, according to Schwadel (1988), typically amount to 35% to 55% of annual retail sales) and other cash flows make December the highest economic activity month and seem to provide part of the reason for January's high returns during the FH and throughout the month for small stocks.

The following explanations for the turn-of-the-month effect have also been advanced: inventory adjustments of different traders (Ritter 1988); the timing of trades by informed and uninformed traders (Admati & Pfleiderer 1988) and specialists' strategies in response to informed traders (Admati & Pfleiderer 1989; Prahl 1993); seasonal tax-induced trading (Lakonishok & Smidt 1986); and window dressing induced by periodic evaluation of portfolio managers (Haugen & Lakonishok 1988; Ritter & Chopra 1989).

The issue of data mining discussed by the authors mentioned above is an important one to consider, especially because, even with strong seasonality effects, the average daily magnitudes of these anomalies are typically less than a two-way transactions cost, which is about 0.46% for NYSE stocks according to Berkowitz *et al.* (1988), or the typical bid-ask spread, which at an eighth is 0.31% of the average-priced $40 NYSE stock.

Perhaps the best remedy against data snooping is new data and convincing reasons for the effects. The reasons for the turn-of-the-month effect are largely cash-flow and institutionally based. Another factor seems to be behavioural. For example, bad news such as that relating to earnings announcements is delayed and announced late in the month, whereas good news is released promptly at the beginning of the month (see Penman 1987).

To investigate this further, Hensel *et al.* (1993) used as a data set the daily closing prices of the Standard & Poor (S & P) 500 Stock Index for the 65-year period from February 1928 to June 1993. This data was supplied by Data Resources Incorporated. The S & P 500 is a value-weighted index of large capitalization US stocks. Since March 1957 this has consisted of the 500 largest stocks weighted by market value (price times number of shares outstanding). Before then it consisted of the 90 largest stocks. This index is called the S & P composite or S & P 500. Index futures contracts on the S & P 500 have been trading since 1982 on the Chicago Mercantile Exchange and futures options are traded at the Chicago Board of Trade. Hensel *et al.* (1994) have studied the turn-of-the month and monthly return patterns of the S & P 500 and the Value Line Composite small stock index during the period of futures trading from 1982 to 1992. In the sequel we will discuss their results, particularly in light of question 5 on futures anticipation. The S & P 500 is an ideal index to study the turn-of-the-month effect because it has a large number of securities (500), is value-weighted (hence each security adds its true weight), is often used as a market portfolio, and is the most widely used passive portfolio benchmark to represent the US equity market in portfolio performance calculations and as an index portfolio. One weakness of this data set is the omission of dividends. Hence it is possible that there is a slight bias in some of the results that follow because of the unevenness of the dividend stream. However, we believe that this effect, if any, is minor, and does not change any of our conclusions.

Figure 1 shows the average return pattern by trading day. The average return on

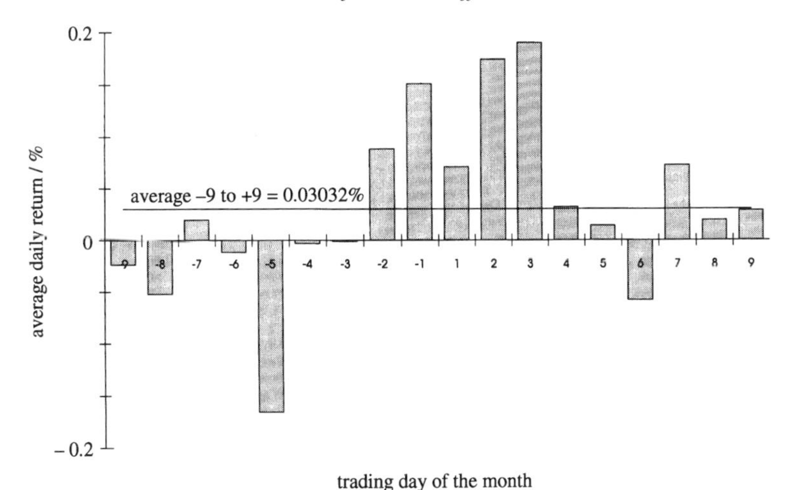

Figure 1. Average daily returns in the S & P 500 cash market by trading day of the month (February 1928 to June 1993). From Hensel *et al.* (1993).

trading days −9 to +9 over the 65-day sample was 0.0303% per day. Days −9 to −1 are during the previous month and +1 to +9 during the current month. For example, day −6 of December is the day in November that is six trading days before the start of December. Significantly higher returns occurred on trading days −2 to +3. Hence, the bulk of the monthly returns occurred at the turn of the month.

Figure 1 suggests that all the monthly gains occurred in the FH. The ROM had negligible returns. The first row of table 1*a* details these mean returns by trading day of the month. The last three columns have the average daily returns in the −9 to +9 period, the TOM (−1 to +4), and the FM (−1 to +9). The mean daily returns over the 65 years were higher in TOM at 0.1236% than in the period −9 to +9, at 0.0303%, with FH having average daily returns of 0.0695%. The first row of table 1*b* gives the *t*-values for the hypothesis that the mean daily returns on the various trading days −9 to +9 were above the mean of the −9 to +9 trading-day returns, which was 0.0303% per day. Trading days −1, +2 and +3 had significantly higher returns than average. Trading days −2 and +1 had high average returns. All of the trading days from −9 to −3 had returns below average. Two of these (−8 and −5) were significantly negative.

Tables 1*a* and 1*b* present the mean returns and *t* statistics, respectively, for the hypothesis that the daily returns were greater than the −9 to +9 average by trading day of the month, by decade from the 1930s to the 1990s and by month. Trading days −1, +2 and +3 had significantly higher mean returns than average in most decades and for the entire 65-year period. Every decade had high returns on −1 and +2, and only in the 1980s were the returns on +3 negative. The latter may be the anticipation of the turn-of-the-month effect in the S & P 500 futures market and the associated index arbitrage; see figure 2 and table 2.

The futures market anticipates the TOM on days −4 to +2, and the index arbitrage keeps the ask prices close to the futures prices. Once the futures index moves on day −4 the expensive futures are sold and cash stocks are purchased for the index

Table 1. (*a*) *Average daily returns in the S & P 500 cash market by decade and month for monthly trading days −9 to +9 (February 1928 to June 1993).* (*b*) *t-values for* (*a*) *and figure 1 for the hypothesis that the mean daily returns are greater than the overall mean for the same time period (for days −9 to +9) by trading day of the month for the S & P 500 cash market by decade and month (February 1928 to June 1993). From Hensel et al. (1993)*

Values significant at the 5% level (two tailed) are in bold. Months are tested against the −9 to +9 mean for the entire period. Specific periods are tested against the mean for the same period.

(*a*)

	−9	−8	−7	−6	−5	−4	−3	−2	−1	+1	+2	+3	+4	+5	+6	+7	+8	+9	−9 to +9	TOM	FH
1928–93	−0.02	−0.05	0.02	−0.01	−0.17	0.00	0.00	0.09	0.15	0.07	0.17	0.19	0.03	0.01	−0.06	0.07	0.02	0.03	0.03	0.12	0.07
1928–39	0.08	−0.21	0.12	−0.01	−0.44	−0.14	−0.19	0.08	0.10	−0.01	0.17	0.40	0.24	0.13	−0.25	0.18	0.16	0.07	0.03	0.18	0.12
1940–49	−0.12	−0.03	0.00	0.00	−0.07	−0.07	−0.02	0.25	0.16	0.09	0.21	0.19	−0.14	0.05	0.03	0.01	−0.16	−0.06	0.02	0.10	0.04
1950–59	0.03	−0.02	0.10	0.05	−0.17	0.11	0.10	0.08	0.16	0.20	0.30	0.18	0.02	0.01	−0.20	−0.02	0.06	0.05	0.06	0.17	0.08
1960–69	−0.08	−0.10	−0.04	−0.05	−0.20	−0.05	−0.03	0.07	0.15	0.04	0.14	0.13	0.10	0.02	0.02	0.11	0.02	0.02	0.01	0.11	0.07
1970–79	−0.12	−0.08	−0.03	−0.08	−0.[illegible]3	0.00	0.08	−0.05	0.09	0.02	0.03	0.21	0.03	0.00	0.01	0.05	−0.01	0.00	0.01	0.07	0.04
1980–93	0.03	0.10	−0.03	0.00	−0.[illegible]7	0.11	0.06	0.09	0.23	0.10	0.19	0.04	−0.06	−0.10	0.04	0.08	0.02	0.08	0.05	0.10	0.06

1928–1993	−9	−8	−7	−6	−5	−4	−3	−2	−1	+1	+2	+3	+4	+5	+6	+7	+8	+9	−9 to +9	TOM	FH
Jan.	−0.03	−0.15	0.10	0.13	−0.06	0.07	0.05	0.14	0.32	0.01	0.56	−0.01	0.19	−0.17	0.10	0.07	0.08	−0.10	0.07	0.21	0.10
Feb.	0.15	−0.13	−0.10	0.08	−0.28	−0.12	0.09	0.01	0.24	0.07	0.15	−0.16	0.07	−0.09	−0.26	0.12	0.00	0.00	−0.01	0.08	0.02
Mar.	−0.05	0.03	−0.03	−0.18	−0.21	0.00	−0.06	0.09	0.16	0.20	0.22	0.14	0.22	−0.04	−0.12	0.17	−0.10	−0.12	0.02	0.19	0.08
Apr.	0.19	−0.10	−0.04	−0.19	−0.22	0.03	−0.14	−0.13	−0.16	0.08	0.15	0.12	0.10	−0.06	−0.02	0.20	0.24	−0.06	0.00	0.05	0.06
May	−0.17	−0.11	0.04	−0.06	−0.11	0.08	−0.07	0.22	0.31	−0.03	0.20	0.32	0.04	0.25	−0.21	0.11	−0.19	0.04	0.04	0.17	0.08
June	−0.04	0.00	0.18	0.02	−0.21	0.01	−0.15	0.08	−0.04	−0.02	0.14	0.31	0.27	−0.02	−0.18	−0.07	0.16	0.15	0.03	0.13	0.07
July	−0.07	−0.10	−0.13	0.23	−0.17	0.14	−0.05	0.10	0.14	0.28	0.32	0.39	−0.01	0.02	0.21	0.08	−0.05	0.31	0.09	0.23	0.17
Aug.	0.03	0.05	0.14	0.07	−0.06	−0.05	−0.01	0.01	0.15	0.11	0.06	0.21	−0.19	0.13	0.03	0.22	−0.08	0.06	0.05	0.06	0.07
Sept.	−0.09	−0.01	0.24	−0.35	−0.25	−0.19	0.06	−0.22	0.25	−0.17	0.17	0.33	−0.09	−0.31	−0.32	0.11	0.02	−0.22	−0.05	0.10	−0.02
Oct.	−0.15	0.19	−0.11	−0.02	−0.49	−0.12	−0.10	0.30	0.00	0.06	0.26	0.18	−0.23	0.47	−0.07	−0.13	−0.26	0.40	0.01	0.04	0.06
Nov.	−0.03	−0.20	−0.14	0.15	0.02	0.00	−0.03	−0.07	0.21	0.25	−0.17	0.34	−0.06	−0.10	0.04	0.08	0.41	−0.05	0.04	0.11	0.09
Dec.	0.03	−0.13	0.13	−0.01	−0.03	0.10	0.33	0.52	0.27	−0.02	0.10	0.15	0.11	0.12	0.04	−0.09	−0.04	−0.09	0.08	0.12	0.06

Table 1 (*cont.*)

(*b*)

	−9	−8	−7	−6	−5	−4	−3	−2	−1	+1	+2	+3	+4	+5	+6	+7	+8	+9	−9 to +9	TOM	FH
1928–93	−1.21	**−1.99**	−0.29	−0.12	**−4.82**	−0.86	−0.78	1.52	**3.31**	1.07	**3.66**	**4.04**	0.04	−0.38	**−2.42**	1.05	−0.27	−0.04	0.00	**5.15**	**3.00**
1928–39	0.26	−1.41	0.69	−0.21	**−3.11**	−1.02	−1.33	0.35	0.50	−0.29	0.86	**2.33**	1.08	0.59	−1.91	0.93	0.88	0.25	0.00	**2.10**	1.77
1940–49	−1.51	−0.74	−0.22	−0.25	−1.18	−0.97	−0.57	**3.07**	**2.22**	0.86	**2.52**	**2.52**	−1.88	0.37	0.12	−0.09	**−2.01**	−0.93	0.00	**2.57**	0.74
1950–59	−0.38	−1.15	0.65	−0.11	**2.50**	0.88	0.72	0.37	1.77	**2.18**	**4.46**	1.68	−0.72	−0.92	**−4.01**	−1.21	−0.04	−0.19	0.00	**4.21**	0.76
1960–69	**−2.07**	**−2.07**	−1.04	−1.13	**−3.88**	−1.14	−0.48	0.84	**2.21**	0.34	**2.00**	**2.28**	1.50	0.11	0.14	1.71	0.16	0.02	0.00	**3.34**	**2.75**
1970–79	**−1.63**	−1.05	−0.50	−1.20	−0.43	−0.12	0.88	−0.69	1.09	0.12	0.35	**2.38**	0.35	−0.11	0.10	0.52	−0.21	−0.12	0.00	1.67	1.28
1980–93	−0.24	0.48	−1.10	−0.69	−1.42	0.93	0.16	0.59	**2.53**	0.67	**2.06**	−0.10	−1.43	−2.01	−0.20	0.41	−0.47	0.38	0.00	1.40	0.47

1928–1993	−9	−8	−7	−6	−5	−4	−3	−2	−1	+1	+2	+3	+4	+5	+6	+7	+8	+9	−9 to +9	TOM	FH
Jan.	−0.59	−1.31	0.91	0.96	−0.59	0.34	0.08	1.06	**3.66**	−0.13	**3.33**	−0.39	0.86	−1.35	0.55	0.42	0.59	−1.09	1.28	**3.03**	1.78
Feb.	0.70	−1.66	−1.00	0.61	**−2.13**	−1.13	0.58	−0.12	**2.31**	0.40	1.26	−1.55	0.46	−1.19	**−2.44**	0.59	−0.20	−0.33	−1.26	1.02	−0.36
Mar.	−0.69	0.03	−0.62	−1.60	**−2.08**	−0.24	−0.63	0.45	1.49	1.85	1.58	1.01	0.73	−0.61	−1.10	1.30	−1.33	−1.22	−0.34	**2.04**	1.16
Apr.	0.81	−0.92	−0.63	**−2.48**	**−2.09**	0.05	−1.38	−1.23	−1.26	0.29	1.02	0.91	0.45	−0.74	−0.19	1.41	1.35	−0.70	−0.77	0.32	0.54
May	−1.48	−0.89	0.22	−0.82	−1.06	0.42	−0.44	1.57	1.80	−0.55	1.49	**2.70**	0.02	1.40	**−2.22**	0.74	**2.11**	−0.07	0.21	**2.34**	1.27
June	−0.68	−0.16	1.28	−0.14	−1.50	−0.16	−1.68	0.32	−0.43	−0.31	0.78	1.53	1.44	−0.60	−1.63	−0.60	0.93	0.71	0.04	1.89	0.97
July	−0.56	−0.74	−1.62	1.28	−1.22	0.75	−0.61	0.60	1.11	**2.11**	**2.55**	**2.83**	−0.33	−0.06	1.85	0.43	−0.80	**2.02**	1.71	**3.58**	**3.96**
Aug.	−0.02	0.12	0.87	0.36	−0.65	−0.70	−0.40	−0.26	**0.95**	0.69	0.21	1.04	−1.89	0.68	0.00	1.28	−0.91	0.26	0.43	0.61	0.74
Sept.	−0.48	−0.28	1.15	**−2.19**	−1.47	−1.58	0.24	−1.79	**2.04**	−1.24	0.98	1.53	−1.31	**−2.00**	**−3.18**	0.67	−0.08	−1.39	**−2.40**	1.12	−0.93
Oct.	−1.01	0.72	−0.80	−0.35	**−2.83**	−0.64	−0.58	1.17	−0.21	0.15	1.92	0.83	−1.21	1.85	−0.62	−0.92	−1.68	**2.18**	−0.62	0.18	0.62
Nov.	−0.36	−1.68	−1.27	0.89	−0.10	−0.18	−0.34	−0.84	1.17	1.48	−0.92	**2.07**	−0.46	−0.71	0.10	0.24	1.78	−0.61	0.21	1.17	1.08
Dec.	0.14	−1.35	1.02	−0.42	−0.51	0.55	**2.44**	**4.50**	**1.64**	−0.39	0.47	0.94	0.78	0.68	0.17	−0.88	−0.56	−1.45	**2.08**	1.51	0.66

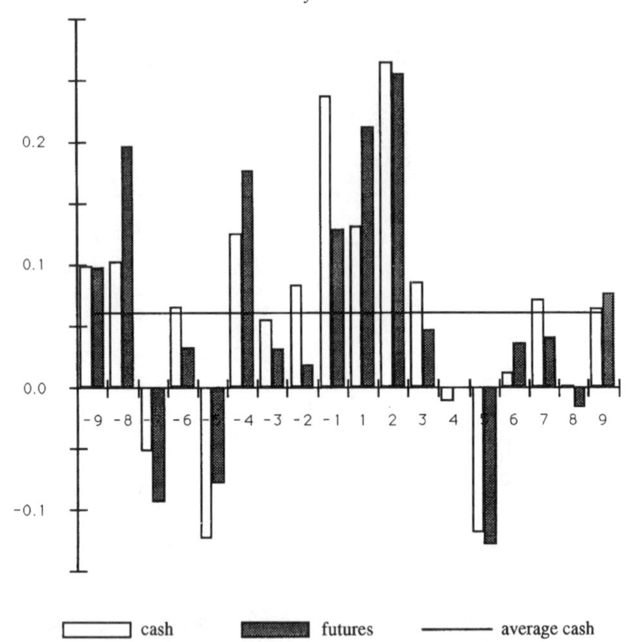

Figure 2. Average daily returns in percent in the cash and futures markets for the S & P 500 by trading day of the month (May 1982 to April 1993) (from Hensel *et al.* 1994).

arbitrage to lock in a nearly riskless profit and increase the level of the stock market. The higher mean returns during TOM and FH and the lower mean returns during ROM were partly an effect of a higher percentage of positive return days as well as a shifting of the conditional means on positive and negative return days. For example, in every decade from 1928 to 1993 the probability of a gain on any of the trading days -2 to $+3$ was always above 50%, and each of these days averaged above 55%. In contrast, trading days in ROM had positive returns on significantly less days which averaged less than 50%. The results are similar for the 65 year sample across months of the year. A table with detailed results appears in Hensel *et al.* (1993). Table 3, which separates the returns in TOM, FH and ROM by month, shows the results clearly. Every month had positive returns in TOM and nearly all had negative returns in ROM. The t-statistics for the hypothesis that the mean returns are greater than average appear in table 4. It is highly significant that the returns in TOM and the FH were above average and those in ROM below average.

There is in excess of $500 billion in equity portfolios indexed to various benchmarks of which by far the largest sum is keyed to the S & P 500. The reason for this is low transactions costs and management fees plus the fact that most managers fail to beat these indices; see Ziemba & Schwartz (1991) for some data across the world. Hence a useful strategy may be to invest in the S & P 500 during TOM or during the FH and be in cash the rest of the time. Table 5 shows the results of these strategies. Both of these strategies mean variance dominated the passive S & P 500 strategy during these 65 years. In particular the risk levels were much lower than those of the S & P 500 and small capitalized stocks.

Table 2. *Average daily returns (log values) in the cash and futures markets for the S & P 500 by trading day of the month, along with t-values for the hypothesis that the day's return differs from the mean return (May 1982 to April 1992). Significant values in bold. From Hensel et al. (1994)*

trading day	average S & P 500 cash daily return (%)	S & P 500 cash t-value	Average S & P 500 futures daily return (%)	S & P 500 futures t-value
−9	0.0992	0.43	0.0973	0.38
−8	0.1025	0.35	0.1959	0.76
−7	−0.0527	−1.28	−0.0932	−1.37
−6	0.0652	0.06	0.0317	−0.28
−5	−0.1239	−1.68	−0.0779	−1.12
−4	0.1246	0.83	0.1757	1.31
−3	0.0548	−0.08	0.0315	−0.31
−2	0.0838	0.27	0.0184	−0.39
−1	0.2361	**2.03**	0.1285	0.72
1	0.1309	0.79	0.2122	1.54
2	0.2637	**2.74**	0.2541	**2.18**
3	0.0852	0.27	0.0467	−0.10
4	−0.0123	−0.79	0.0000	−0.53
5	−0.1181	−1.95	−0.1276	−1.61
6	0.0124	−0.61	0.0354	−0.25
7	0.0716	0.13	0.0410	−0.16
8	0.0003	−0.69	−0.0163	−0.72
9	0.0647	0.05	0.0759	0.18
average −9 to 9	0.0604	—	0.0572	—

Table 3. *Average daily returns in the S & P 500 cash market by month, during* TOM, FH, ROM *and whole month (February 1928 to June 1993). From Hensel et al. (1993)*

S & P 500 index	TOM −1 to +4	FH −1 to +9	ROM +10 to −2	all days −1 to −1
Jan.	0.2061	0.1025	0.0359	0.0651
Feb.	0.0807	0.0170	−0.0214	−0.0024
Mar.	0.1876	0.0768	−0.0212	0.0208
Apr.	0.0503	0.0566	−0.0169	0.0161
May	0.1653	0.0819	−0.0836	−0.0107
June	0.1287	0.0669	0.0033	0.0315
July	0.2258	0.1697	−0.0050	0.0738
Aug.	0.0645	0.0672	0.0129	0.0364
Sept.	0.0976	−0.0175	−0.0978	−0.0605
Oct.	0.0445	0.0632	−0.0787	−0.1780
Nov.	0.1108	0.1038	−0.0821	0.0071
Dec.	0.1217	0.0564	0.0599	0.0584
all months	0.1236	0.0703	−0.0235	0.0186
all except Jan.	0.1162	0.0674	−0.0292	0.0143

The results in table 5 are suggestive. But the reader should note the following: (*a*) the S & P 500 results include dividends; (*b*) the small capitalized index includes dividends and is the value-weighted average of the bottom 20 % of NYSE stocks; (*c*) the intermediate-term government bonds are total return for five to 10 year maturities; (*d*) the TOM and FH plus cash returns do not include dividends, hence the mean returns are biased lower by an amount that is probably greater than one might

Table 4. *t-values for the hypothesis that the mean returns are greater than average in the S & P 500 cash market by month, during* TOM, FH, ROM *and whole month (1928–1993). From Hensel et al. (1993)*

S & P 500 index	TOM −1 to +4	FH −1 to +9	ROM +10 to −2	all days −1 to −1
Jan.	3.11	2.08	0.56	1.87
Feb.	1.31	−0.04	−0.96	−0.77
Mar.	2.61	1.41	−1.22	0.09
Apr.	0.57	0.93	−0.95	−0.09
May	2.84	1.65	−2.56	−1.04
June	1.55	1.02	−0.38	0.42
July	3.96	4.16	−0.59	2.00
Aug.	0.80	1.18	−0.15	0.65
Sept.	1.22	−0.77	−2.33	−2.29
Oct.	0.38	0.84	−1.62	−0.88
Nov.	1.20	1.53	−2.41	−0.33
Dec.	1.79	0.95	1.18	1.52
all months	5.94	4.13	−3.71	0.00
all except Jan.	5.28	3.54	−3.76	−0.47

Table 5. *Comparison of the* TOM *and* FH *plus cash strategies with S & P 500 small capitalized stocks and intermediate term government bonds (1928–1993). From Hensel et al. (1993)*

	yearly average returns (%)	yearly std. dev. of returns (%)
S & P 500	9.5	20.1
Sm. Cap stocks	11.5	30.1
It Gov. bonds	5.1	4.4
TOM (−1 to +4)+0.8 cash	10.1	8.8
FH (−1 to +9)+0.6 cash	11.1	12.6

expect to spend on transactions costs and by market impact using futures trading to implement these strategies.

3. The turn-of-the-month effect in worldwide equity markets

The turn-of-the-month effect seems to occur in most large equity markets around the world. Ziemba (1989, 1991; see also the detailed summary in Ziemba & Schwartz 1991) studied the effect in Japan. By using data on the Nikkei 225 Index for the years 1949–1988, he found that the turn of the month is similar in Japan except that the dates change, with the turn being −5 to +2 and with +3 to +7 being the rest of the first half of the month. All the days −5 to +2 had significantly positive returns, as shown in table 6. The FH trading days −5 to +7 had mean returns of 0.01142% per day against 0.00093 for the ROM days +8 to −4. Hence, essentially all the gains were in the FH.

The reasons for the effect in Japan during this period were as follows and added credence to the existence of the effect.

(i) Most salaries were paid between the 20th and the 25th of the month, with the 25th being especially popular.

(ii) There was portfolio window dressing on day −1.

(iii) Security firms could invest for their own accounts based on their capitalization. Since their capitalization usually rose each month and was computed at

Table 6. *Mean returns (%) on the* NSA *by trading day of the month 1949–1988. From Ziemba (1991)*

trading day	sample size	mean return (%)	t-statistic	trading day	sample size	mean return (%)	t-statistic
−5	471	0.0899	2.75	10	471	0.062	1.58
−4	471	0.1041	2.65	11	471	0.0395	1.05
−3	471	0.1733	3.82	12	471	−0.0196	−0.50
−2	471	0.1334	3.18	13	471	0.0115	0.24
−1	471	0.2255	5.36	14	471	−0.0042	−0.10
1	471	0.0980	2.43	15	471	−0.0306	−0.54
2	471	0.1006	2.47	16	471	0.0716	1.35
3	471	0.0307	0.72	17	471	−0.0498	−1.20
4	471	0.0592	1.15	18	429	−0.0207	−0.49
5	471	0.0358	0.81	19	350	0.0162	0.36
6	471	−0.0005	−0.01	20	229	−0.0286	−0.48
7	471	0.0357	0.09	21	118	−0.0476	−0.46
8	471	−0.0585	−1.34	22	39	0.0562	0.45
9	471	0.1065	2.46				

the end of the month, there was buying on day −3 to account for this. Buying was done as soon as possible.

(iv) Large brokerage firms had a sales push that lasted 7–10 days starting on day −3.

(v) Employment stock-holding plans and mutual funds received money to invest in this period starting around day −3.

(vi) Individual investors buy mutual funds with their pay, which they receive on calendar days 15th–25th of the month; the funds were then invested in stocks with a lag, so most of the buying occurred on days −5 to +2.

(vii) For low liquidity stocks, buying occurred over several days by dealing in accounts to minimize price-pressure effects.

Ziemba (1989) investigated the futures market trading outside Japan on the SIMEX in Singapore during the turn of the month and examined other anomalous effects in Japanese security markets during the period of September 1986 to September 1988 before there was futures trading on the NSA or TOPIX in Japan. The results were that the spot effect was consistent with past data, so the futures market did not alter the effect.

However, the futures market in Singapore totally anticipated the effect on days −8 to −5 with a total average rise of 2.8%; see figure 3. Then, when the effect occurred on days −5 to +2 and the spot market gained 1.7%, the futures market was flat.

The turn-of-the-month effect existed and was strong in Japan from 1949 to 1988. However, the extreme decline in the Japanese stock market of over 60% from January 1990 to August 1992 so shook investor confidence that there was no confidence in equity investments. Hence, one would not expect that the usual cash flows would enter the market in the usual way to generate the TOM and FH high returns. Indeed this was the case, and returns during these periods were not higher than average during 1990–1993. See Bell & Ziemba (1994) for an analysis of Japanese anomalies during 1990–1993 and Stone & Ziemba (1993) for an analysis of this period in Japanese economic history.

Cadsby & Ratner (1991) investigated turn-of-the-month and holiday effects in the US, Canada, the UK, Australia, Switzerland, West Germany, Japan, Hong Kong,

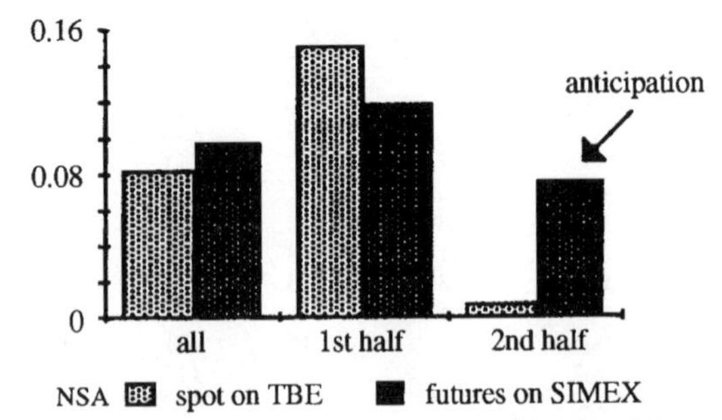

Figure 3. Daily rates of return in percent in the first and second halves of the month in Japan. Cash data, futures market effects September 1986 to September 1988 (from Ziemba 1989).

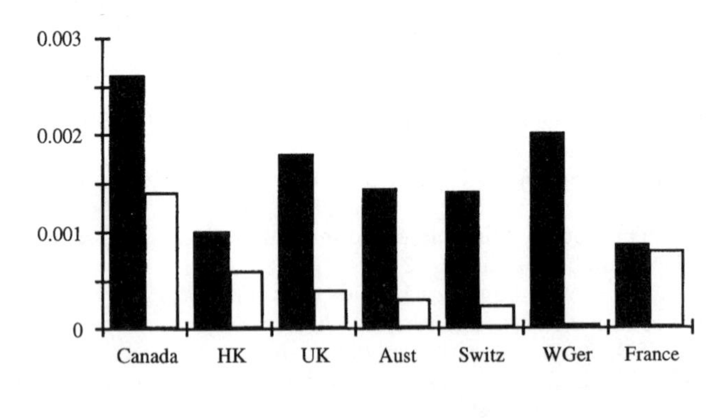

Figure 4. International TOM effects. Daily rates of return in percent for Canada (1/75–12/87), Hong Kong (1/80–1/89), U.K. (8/83–6/88), Australia, Switzerland, West Germany and France (all 1/80–8/89) (from Cadsby & Ratner 1991).

Italy and France. Their use of -1 to $+4$ as the turn of the month in all these countries makes their results hard to interpret since there may be cash flow and institutional constraint date alterations as in Japan. Still they found a significant turn-of-the-month effect in most countries, as shown in figure 4. A good example of a significant turn-of-the-month effect with different dates is that in Italy. The month there essentially starts on the 15th (approximately) and all trades from then up to the following 14th are settled on the last trading date of that calendar month. This system leads to high returns from the 15th to the end of that calendar month and low or negative returns from the $+1$ day to the 14th; see Barone (1990). Not suprisingly, Italy has not had high returns on days -1 to $+4$ and does not appear in figure 4.

4. Concluding remarks

The turn-of-the-month effect seems to be a worldwide phenomenon. The paper by Martikainen *et al.* (1994), although it only considers the brief period from January 1988 to January 1990, shows this with various indices for a variety of regions of the world. They also found that returns on individual days during TOM, especially -1, were high. The latter averaged 0.41 % per day over 24 countries with a t value of 3.89 and was the only trading day with a mean return different from zero.

The TOM, FH and ROM results are useful for traders and portfolio managers. However, points 1–7 mentioned earlier in this paper must always be considered. One of us (W. T. Ziemba) has used such ideas in several investment accounts with some success. Although there was co-mingling of various anomalous ideas such as those discussed in Ziemba (1994*b*), the ideas discussed here were useful inputs to a diversified strategy that returned 27.0 % per year net of all costs for the 77 months from October 1987 to February 1994, with a yearly standard deviation of 8.0 % during 1991–1994.

Research was partly supported by the Social Sciences and Humanities Research Council of Canada and Frank Russell Company.

References

Admati, A. R. & Pfleiderer, P. 1988 A theory of intra-day patterns: volume and price variability. *Rev. Financial Stud.* **1**, 3–40.

Admati, A. R. & Pfleiderer, P. 1989 Divide and conquer: a theory of intra-day and day of the week mean effects. *Rev. Financial Stud.* **2**, 189–223.

Ariel, R. A. 1987 A monthly effect in stock returns. *J. Financial Econ.* **9**, 3–18.

Barone, E. 1990 The Italian stock market: efficiency and calendar anomalies. Centre for Research in Finance, IMI Group. (Working paper.)

Banz, W. 1981 The relationship between return and market value of common stocks. *J. Financial Econ.* **9**, 3–18.

Bell, D. & Ziemba, W. T. 1994 Investigation of seasonal anomalies on the Tokyo stock exchange, 1990–93. University of British Columbia. (Working paper.)

Berkowitz, S. A., Logue, D. E. & Noser, E. A. 1988 The total cost of transactions on the NYSE. *J. Finance* **43**, 97–112.

Black, F. 1986 Noise. *J. Finance* **41**, 10–14.

Black, F. 1992 Estimating expected returns. Goldman Sachs, New York. (Working paper.)

Black, F., Jensen, M. & Scholes, M. 1972 The capital asset pricing model: some empirical tests. In *Studies in the theory of capital markets* (ed. M. Jensen). New York: Praeger Publishers.

Blume, M. E. & Siegel, J. J. 1992 The theory of security pricing and market structure. *J. Financial Markets, Institutions & Instruments* **1**, 1–58.

Blume, M. E. & Stambaugh, R. F. 1983 Biases in computed returns: an application to the size effect. *J. Financial Econ.* **12**, 387–404.

Cadsby, C. B. 1992 The CAPM and the calendar: empirical anomalies and the risk-return relationship. *Man. Sci.* **38**, 153–1561.

Cadsby, C. B. & Ratner, M. 1991 Turn-of-the-month and pre-holiday effects on stock returns: some international evidence. University of Guelph. (Working paper.)

Clark, R. & Ziemba, W. T. 1987 Playing the turn-of-the-year effect with index futures. *Operations Res.* **35**, 799–813.

Cootner, P. (ed.) 1964 *The random character of stock market prices*. Cambridge, Mass.: MIT Press.

Dimson, E. (ed.) 1988 *Stock market regularities*. Cambridge University Press.

Fama, E. F. 1965 The behaviour of stock market prices. *J. Business* **38**, 34–105.

Fama, E. F. 1970 Efficient capital markets: a review of theory and empirical work. *J. Finance* **25**, 383–417.

Fama, E. F. 1991 Efficient capital markets: II. *J. Finance* **46**, 1575–1617.

Fosback, N. 1976 *Stock market logic*. Fort Lauderdale, Florida: Institute for Economic Research.

Haugen, R. A. & Lakonishok, J. 1988 *The incredible January effect*. Homewood, Illinois: Dow Jones–Irwin.

Hawawini, G. & Keim, D. 1994 On the predictability of common stock returns: world-wide evidence. In *Finance* (ed. R. Jarrow, V. Maksimovic & W. T. Ziemba). Amsterdam: North-Holland.

Hensel, C., Sick, G. A. & Ziemba, W. T. 1993 The turn-of-the-month effect in the S & P 500, 1926–93. University of British Columbia. (Working paper.)

Hensel, C., Sick, G. A. & Ziemba, W. T. 1994 The turn-of-the-month effect in the US index futures markets, 1982–92. *Rev. Futures Markets.*

Hirsh, Y. 1986 *Don't sell stocks on Monday.* New York: Facts on File Publications.

Jaffe, J. D., Keim, B. & Westerfield, R. 1989 Earnings yields, market values and stock returns. *J. Finance* **44**, 135–148.

Keim, D. B. 1983 Size related anomalies and stock return seasonality. *J. Financial Econ.* **12**, 13–32.

Kendall, M. G. 1953 The analysis of economic time-series. Part I: Prices. *Jl R. statist. Soc.* **96**, 11–25.

Lakonishok, J. & Smidt, S. 1986 Volume for winners and losers: taxation and other motives for stock trading. *J. Finance* **41**, 951–974.

Lakonishok, J. & Smidt, S. 1988 Are seasonal anomalies real? A ninety-year perspective. *Rev. Financial Stud.* **1**, 403–425.

Lo, A. W. & MacKinlay, A. C. 1990 Data-snooping biases in tests of financial asset pricing models. *Rev. of Financial Stud.* **3**, 431–467.

Mandelbrot, B. 1963 The variation of certain speculative prices. *J. Business* **36**, 394–419.

Markowitz, H. 1992 Data mining. Daiwa Securities, Jersey City, New Jersey. (Working paper.)

Martikainen, T., Perttunen, J. & Ziemba, W. T. 1994 Finanzmarkt und Portfolio Management. (In the press.)

Merton, R. C. 1985 On the current state of stock market rationality hypothesis. In *Macroeconomics and Finance* (ed. R. Dornbusch & S. Fischer). Cambridge, Mass.: MIT Press.

Ogden, J. P. 1987 The end-of-the-month as a preferred habitat: a test of operational efficiency in the money market. *J. Financial Quantitative Anal.* **22**, 329–344.

Ogden, J. P. 1990 Turn-of-the-month evaluations of liquid profits and stock returns: a common explanation for the monthly and January effects. *J. Finance* **45**, 1259–1272.

Penman, S. H. 1987 The distribution of earnings news over time and seasonalities in aggregate stock returns. *J. Financial Econ.* **18**, 199–228.

Prahl, W. 1993 Stock return seasonality and market maker monopoly power. University of Wisconsin, Madison. (Working paper.)

Reinganum, M. R. 1981 Misspecification of capital asset pricing: empirical anomalies based on earnings, yields and market values *J. Financial Econ.* **9**, 19–46.

Ritter, J. R. 1988 The buying and selling behavior of individual investors at the turn-of-the-year. *J. Finance* **43**, 701–717.

Ritter, J. R. & Chopra, N. 1989 Portfolio rebalancing and the turn-of-the-year effect. *J. Finance* **44**, 149–166.

Roberts, H. V. 1959 Stock market patterns and financial analysis: methodological suggestions. *J. Finance* **14**, 1–10.

Roberts, H. V. 1967 Statistical versus clinical prediction of the stock market. University of Chicago. (Working paper.)

Roll, R. 1977 A critique of the asset pricing theory's tests; part I: on past and potential testability of the theory. *J. Financial Econ.* **4**, 129–176.

Schwadel, F. 1988 Retailers see a not-so-ho-ho Christmas. *Wall Street J.*, **70**, A2 (11 November).

Stone, D. & Ziemba, W. T. 1993 Land and stock prices in Japan. *J. Econ. Perspectives*, 149–165. (Summer.)

Thaler, R. H. 1992 *The Winner's Curse.* New York: The Free Press.

Ziemba, W. T. 1989 Seasonality effects in Japanese futures markets. In *Research on Pacific Basin Security Markets* (ed. Rhee & Chang), pp. 379–407. Amsterdam: North-Holland.

Ziemba, W. T. 1991 Japanese security market regularities: monthly, turn-of-the-month and year, holiday and Golden Week effects. *Japan World Econ.* **3**, 119–146.

Ziemba, W. T. 1994*a* *Strategies for making and keeping excess profits in the stock market*. New York: William Morrow. (In the press.)

Ziemba, W. T. 1994*b* Worldwide security market regularities. *European J. Operational Res.* (In the press.)

Ziemba, W. T. 1994*c* Investing in the turn-of-the-year effect in the futures markets. *Interfaces*. (In the press.)

Ziemba, W. T. & Schwartz, S. L. 1991 *Invest Japan: the structure, performance and opportunities of the stock, bond and fund markets*. Chicago: Probus Publishing.

6

Making money from mathematical models

David Harding

This paper is a discussion of the scientific significance and nature of mathematical models generally, and in finance theory in particular; of the relation of such models to a postulated 'reality'; of the sufficiency of the empirical grounding of such models to the task of drawing useful inferences about such a reality; and of the danger of unwittingly propagating erroneous conclusions about this reality in the absence of such a grounding. Some suggestion as to how the current empirical grounding of mathematical modelling in finance can be enriched.

1. Introduction

There is a little remarked upon dissonance between the philosophy of mathematics and that of the natural sciences. Mathematics is an abstract philosophy with an essentially aesthetic nature. Many mathematicians and philosophers have asked whether it is best to describe mathematics as having been discovered or invented. Perhaps more pragmatic mathematicians would be inclined to follow Wittgenstein's advice and pass over in silence that of which they cannot speak! The choice between these options is one with no profound consequences for the practice of mathematics.

Natural science has no such incipient philosophical argument at its heart. The philosophy of natural science may not be generally well understood but it is, to my knowledge, relatively uncontentious.

The natural scientist observes naturally occurring phenomena and attempts first to classify and subsequently to model them (in a theory) by a process of metaphor and analogy. If such a description seems at first puzzling, we must reflect on the logic of the fact that the theory cannot be the reality; it can at best only be a near perfect simulation of that reality. At any time the reality to be modelled consists of the universe of recorded observations, a universe which as long as mankind continues to progress expands so that even a model in accordance with every known observation could not properly be described as 'true' in the sense that it is necessarily identical to the postulated reality. The discipline which, in my opinion, has made the philosophy of natural science so productive for mankind is the agreement that such a model must be capable of producing empirically falsifiable predictions and must thus be capable of being subjected to an unbiased test of its value within its own conceptual framework. A theory's utility is then determined by the quality of the predictions it makes and the accuracy with which these accord with empirical observation. No natural scientist can ever really be excused, however, of believing or stating that such and such a theory is 'true' according to the most puritanical interpretation of that word, or 'represents reality' or other such sentiments. It is utility that is the distinguishing touchstone of the philosophy of natural science in its competition with other philosophies.

Mathematical Models in Finance. Edited by S.D. Howison, F.P. Kelly and P. Wilmott.
Published in 1995 by Chapman & Hall, London. ISBN 0 412 63070 2

The mathematician does not, by necessity, share this philosophical basis with the natural scientist, and yet the startling practical successes of natural science over the past 300 years have been achieved by generations of natural scientists working with the 'armoury' provided by mathematicians and still being vigorously expanded today. The natural scientists have 'subverted' mathematics for their own modest but determined aims: to construct models with the greatest and richest power to illuminate and draw inferences about the nature of postulated reality.

The mathematician's criterion of virtue is by contrast an aesthetic one and when working with the natural scientist this should be emphasized. We may perceive reality as beautiful and mathematics as beautiful but it would not be logically correct to infer that mathematics is reality, that the elegant solution is necessarily the best.

This should be commonplace among natural scientists, but the success of the scientific method over the past 300 years is such that these essential philosophical foundations are insufficiently well remembered. A particularly grotesque mutilation of scientific philosophy is performed by those who claim that particular scientific theories are true 'to all intents and purposes'. Newtonian mechanics explained the movements in the heavens to a degree sufficient for all practical purposes. However, quantitatively speaking the tiniest chink in the accordance of the theory's prediction with reality was sufficient to open the way for the einsteinian revolution. The discovery of such a chink had to await the considerable technological advances in optics and instrumentation of the nineteenth century before it could be perceived. The atomic bomb is not a negligible consequence of the tiniest imperfection in a 'nearly true' theory. Out of a tiny inconsistency between observed reality and an aesthetically complete theory arose not just enormous practical consequences, but a revolution in our understanding of the relation between the natural sciences and the reality that is perceived as their subject. In science, as in other areas of life, ladders must be climbed only so as to be kicked away. The father of natural science was Plato, whose notion of an absolute reality capable of limitless investigation through reason has inspired centuries of effort to elucidate this reality. Yet Heisenberg's, as yet unrefuted, theory is arguably consistent with the idea that such an absolute reality can never be observed. The idea that models are nearly true or are true for all practical purposes must be rejected as forcefully as the idea that they are true. If they look true we must look for the explanation. For them to be true in the strictest sense demands the impossibility of an observation out of accordance with them; a position unworthy of further discussion.

It is the empirical success of natural scientific philosophy more than any other body of ideas that has made the modern world much more than a mere continuation of the ancient. The industrial revolution is Great Britain would have been impossible without the spread of natural scientific philosophy in the 17th and 18th centuries. In a letter to Adam Smith commenting upon *The Wealth of Nations*, dated 10 September 1759 Edmund Burke wrote, 'A theory like yours, founded on the nature of man, which is always the same, will last, when those that are founded upon his opinions, which are always changing, will and must be forgotten.'

Natural science provides an independent method of arbitration between views, and this more than any other factor has enabled the replacement of sterile conflict and assertion with constructive argument and benign progress.

On the human timescale, however, we must not fail to appreciate that the philosophy of natural science is still new. Political, theological and economic organization and activity have thrived through the long ages of history under the sway of less modest philosophies and where political and theological ideologies or economic interest have collided with the timorous ambitions of the natural scientist, it is often the latter who, in the short term, has had to yield. These philosophies and ideologies influenced by the competitive successes of the natural sciences have sought the company of mathematics, perhaps in an attempt to bathe in the reflected light of reason. But mathematics has not prospered in such company as it has with the philosophy of natural science, which has provided it simultaneously with a raison d'être and a continual source of fresh stimuli. Mathematics does not need natural science, but it thrives on its company.

2. Scientific method and modelling market behaviour

Now let us apply these thoughts, slightly randomly, to developments in the mathematical modelling of market behaviour. I have already referred to Adam Smith who was among the first to formalize a model of the growth of economic and social systems founded on the philosophy of natural science. For those who know of market economics only by ill-repute it may perhaps come as a surprise to learn that Smith's mode of construction, as Burke's words testify, was deeply empirical, founded on detailed and analytical observation of the nature of human society. From this construction, the establishment of freely determined market prices in both agricultural and manufactured goods emerges as the optimum method for establishing values where the aim is to foster the maximum rate of that quantity, economic growth, which is seen as a natural consequence of a society in which people are free to interact economically and when the rule of law and the defence of property are guaranteed. This model, its laws of supply and demand and its concept of the division of labour have played the role slightly akin to that Newton's laws of motion played in physics in the development of the science of economics since. But although Newton's ideas were taken up to extraordinary effect by those studying a nature that does not talk back, in economics as in other social sciences, progress has been held back by the sheer practical difficulty of employing the experimental method. It has rarely been possible to experiment in a controlled fashion with society as a whole, and when something resembling experimental conditions are created, the 'adjudicator' of natural science is unlikely always to be respected by the larger part of the jury. Thus, today rather than being seen as the 'progenitor' of economic modelling, in the way that Stephenson is the 'progenitor' of the steam train or Brunel the 'progenitor' of civil engineering, Smith is seen as a 'political' figure and his excellent, and in my opinion, proven, contribution, scientifically undervalued. This is not to take sides in the modern political debate which is often concerned with multidimensional problems beyond the scope of Smith's work but is merely to observe that Smith's models did give rise to empirically testable, interesting and often counter-intuitive predictions about the world and which in the messy social laboratory of the past three centuries have received far more support than contradiction. That the experiments took a long time to perform, that they are always necessarily less conclusive than controlled experiments and that any analysis made of them will not necessarily be accepted by social scientists are reasons progress in the field of social and political science has been so much less impressive than in the physical or biological sciences. In the physical sciences the object of study does not have a mind of its own!

I mention Adam Smith because his models were constructed long enough ago for them to have been tested for robustness under a wide range of conditions. The evolution of social systems proceeds at a slower pace than that at which we are nowadays accustomed to living and thus it is much more realistic to judge the usefulness of models developed some time ago, without prior knowledge, than it is to assess their possible contemporary equivalents. Had Smith lived through the subsequent centuries there would have been many occasions on which he would have been able to profit handsomely by exploiting the difference between his theories' predictions and the common opinion and that is the ultimate judge of the utility of his model, of its value as a scientific theory.

I have stressed that a theory or model is scientifically valueless unless it is capable of making empirically falsifiable predictions. In economics and the study of markets there can be no more incorruptible measure of utility than profit accrued by arbitraging the predictions of a model against the common view (the market). The quest to 'beat the market' is thus more than a venal desire for money or an egotistical desire to win, it is also a means of obtaining the most ruthlessly honest evaluation of the scientific utility of a model or method.

Speculari, the Latin root of the verb to speculate, has the literal meaning 'to observe'. And a study of speculation will show that most successful speculators can be well described as 'observers'. To be successful, this observation must of necessity be detached and unemotive and thus, where great social and moral issues are at stake, it is perhaps not surprising that this viewpoint should arouse some distrust and hostility among the general population (particularly when the speculator profits at a time of general discontent). Yet this detached observation is clearly in the spirit of the natural scientist and the act of speculating for money is in the spirit of the empirical scientist's restless yearning to add to empirical knowledge and put theories to the test. Thus, making money from mathematical models is in one sense less about the corruption of intellectual endeavour than about the appropriate statistical test of the utility of such models for the development of scientific theory.

3. The utility of the efficient market theory

It is in this context that I wish to consider the scientific utility of the 'efficient market theory'. The various versions of the theory begin essentially by asserting that it is impossible to make money by applying mathematical modelling to the science of speculation. What, then, is its utility as a scientific theory? On the one hand its predictions of market price are of the null variety – that no better estimate of tomorrow's price than today's can be discerned – and not very interesting. On the other hand, a concrete prediction that future returns will be drawn from a known distribution whose parameters can be estimated appears falsified; the empirical evidence points to the return process in all markets being ultimately non-parametric and certainly non-stationary. Its great strength is that it is consistent with one of the most profoundly useful insights about market behaviour: it is very difficult to make money consistently. Such consistency, however, it not a unique feature of this model over a universe of alternatives.

What of the practical evidence? Because every major bank and securities house now has its option software and its rocket scientists surely they must be making money from the models thus indirectly confirming their utility. But how is this money made? First through arbitrage – using the model to assess the relative value of various forms of derivatives of the same asset or assets – a test that is relatively insensitive to the crucial distributional assumptions underlying the theory; second, through what we may (not necessarily derogatorily) call merchandising: banks and brokers selling at marked up prices derivative instruments that can only be created because of the existence of the theory. These profits do not ultimately refute the theories' scientific utility. If this sounds contrived consider the case of portfolio insurance. Some made personal fortunes from selling advice based on the theory (in good conscience) but to compensate, after the market crash of 1987, their pension fund clients incurred losses greater than they otherwise would have done. Thus the widespread use, found for the theory, is not strong evidence of its scientific utility but more for its marketability. None of this is to denigrate the contribution of the theory towards improved practice in and greater understanding of investment but it is to point out sharply its limited ambition and limited utility as a scientific theory and to undermine the perception of confirmation its widespread usage suggests. It is no surprise of course that speculators should be in conflict with the theory because it explicitly denies the possibility of their existence.

4. Conclusion

Having aired my doubts I now have some positive comments on how to make money from mathematical models or perhaps more properly how I have observed money being made. I believe there are three distinct paths that can be followed.

1. Be a purveyor of derivative instruments or shareholder in such an activity. This has all the intellectual purity of selling vegetables!

2. Be a rocket scientist arbitrageur. The efficient market theory is sufficiently robust with respect to relative values and some very challenging mathematics has been required to unlock new arbitrage potential in the globalizing financial markets. This can be challenging and satisfying but its assumptions may be unsound.

3. Speculate, which I believe to be the intellectual front line. One may study and observe the world so as to seek phenomena amenable to classification and to form ideas as to the metaphors and analogies that are components of a model that can make interesting and falsifiable predictions.

The disincentive to a mathematician of pursuing course 3 is that observation and classification are not the mathematician's job! But any mathematician motivated by the philosophy of science will not find a shortage of opportunity in this course. Efficient market theory has at least partly driven the charlatan from the investment stage. Perhaps this has created the opportunity for the scientist to take to that stage and to push further back the frontier of ignorance for the betterment of humankind.

7

Path-dependent options and transaction costs

J.N. DEWYNNE, A.E. WHALLEY AND P. WILMOTT

We describe research in the subjects of exotic option pricing and option pricing when trade in the underlying incurs transaction costs. These two subjects are then formally brought together to model, in terms of differential equations, problems in pricing exotic options with transaction costs. Results are presented in several cases.

1. Introduction

There has recently been a great deal of interest both in the academic literature and among practitioners in the two subjects of option pricing in the presence of transaction costs and valuing exotic options. As we shall shortly see, valuing options when there are costs associated with trading the underlying leads to some interesting mathematical problems with important consequences. With the current popularity of exotic options it is clearly important to be able to price such products in the same framework. As exotic options become more frequently traded so profit margins will be squeezed and the trader with the most accurate models will make the most profit. Part of this accurate modelling is knowing the effect of transaction costs on an option's price.

In §2 of this paper we describe the modelling of exotic option prices as the solution of partial differential equations. We set up a very general framework into which we can incorporate many path-dependent options including Asians and lookbacks. We derive the relevant partial differential equations along with boundary, final and jump conditions and constraints.

In §3 we describe in detail the generalized Leland model for the effect of transaction costs on vanilla options. We then review other partial differential equation models for other trading strategies. We assume throughout a fairly general transaction cost structure.

In §4 we put together the results of §§2 and 3 in a partial differential equation model for pricing exotic options with transaction costs. We briefly discuss the possibility of similarity solution and the effects of discrete sampling of the path-dependent quantities.

Some results of numerical simulations for discretely sampled lookback strike puts are presented in §5.

2. Pricing exotic options

(a) *Some common exotics*

The value of an option depends on its pay-off at exercise or expiry. In the case of vanilla calls this pay-off is simply $\max(S-E,0)$, where S is the price of the

Mathematical Models in Finance. Edited by S.D. Howison, F.P. Kelly and P. Wilmott.
Published in 1995 by Chapman & Hall, London. ISBN 0 412 63070 2

underlying asset and E is the exercise price, a given constant. From this most basic of option contracts it is a small step to the binary option. Such an option has a more general pay-off, for example $V(S,T) = \mathscr{H}(S-E)$, where V is the price of the option, T is the expiry date and $\mathscr{H}(\cdot)$ is the Heaviside function. (This can be interpreted as a straight 'bet' on whether the asset price will be above or below E at expiry.) The next stage in increasing complexity is the barrier option. A simple example of a barrier option is a 'down-and-out'. This is similar to vanilla option but with the extra specification in the contract that should S ever fall to a prescribed level X, say, before expiry, then the option becomes worthless.

American options (which permit early exercise, as opposed to European options which do not) and barrier options are 'path-dependent': the value depends on the realized asset price path. However, this dependency is rather simple and it is still possible to write the option price as a function of S and t only, $V(S,t)$. Examples of option contracts that are path-dependent in a non-trivial manner are Asian options and lookback options. The former option has a pay-off depending on an average of the realized asset price random walk and the latter on the realized maximum (or minimum).

In saying that the Asian option pay-off depends on the average of the asset price we must be very precise in our definition of 'average'. Two obvious definitions exist for an average (and there are many more less obvious ones): the arithmetic average and the geometric average. These two types of average are distinguished by whether the mean is taken of the asset prices or the logarithm of the asset prices. Finally, the average may be continuous or discrete, depending on whether all asset prices are used in the mean or only those at prescribed 'sampling dates'. The lookback option may similarly depend on either the continuously sampled or the discretely sampled maximum (or minimum).

(*b*) *The general framework*

The options above can all be put into the basic Black–Scholes framework (Black & Scholes 1973) with very little effort.

Let us suppose that the option contract specifies a pay-off which is a function of S and an integral over the underlying's realized history of the form

$$I = \int_0^t f(S(\tau), \tau)\, d\tau.$$

Thus the pay-off at exercise has the form $\Lambda(S, I, t)$. For example, when

$$f = S,$$

this gives a dependence on the continuously measured arithmetic average. When

$$f = \log S,$$

we get the continuously sampled geometric average. When

$$f = S \sum_{i=1}^{N} \delta(t - t_i),$$

we have the discretely sampled arithmetic average with N sampling dates t_i. When

$$f = \log S \sum_{i=1}^{N} \delta(t - t_i),$$

we have the discretely sampled geometric average and, finally, when

$$f = S^n,$$

(multiplied by delta functions if necessary) and

$$J = \left(\int_0^t f(\tau)\,\mathrm{d}\tau \right)^{1/n},$$

we get, in the limit $n \to \infty$, the maximum of the asset price, or as $n \to -\infty$, the minimum. (Thus both Asian options and lookback options have pay-offs which depend on I or J, provided f is suitably defined.)

Following the derivation of the original Black–Scholes equation we assume that S follows the random walk given by the geometric brownian motion

$$\mathrm{d}S = \mu S\,\mathrm{d}t + \sigma S\,\mathrm{d}X,$$

where $\mathrm{d}X$ is a normally distributed random process with $E(\mathrm{d}X) = 0, E(\mathrm{d}X^2) = \mathrm{d}t$. We set up a portfolio of one option, with value $V(S, I, t)$, and $-\Delta$ of the underlying asset. If Π is the value of this portfolio then

$$\Pi = V - \Delta S.$$

From Itô's lemma we have

$$\mathrm{d}\Pi = \mathrm{d}V - \Delta\,\mathrm{d}S = (V_t + fV_1 + \tfrac{1}{2}\sigma^2 S^2 V_{SS} - \Delta D(S, I, t))\,\mathrm{d}t + (V_S - \Delta)\,\mathrm{d}S,$$

where D is the dividend on the underlying asset: an amount $D(S, I, t)\,\mathrm{d}t$ is paid on the asset from t to $t + \mathrm{d}t$. We choose $\Delta = V_S$ to eliminate the random terms. This leaves a deterministic portfolio which, in the absence of early exercise, must have a return equal to that from a risk-free deposit. Thus

$$\mathscr{L}(V) = V_t + fV_I + \tfrac{1}{2}\sigma^2 S^2 V_{SS} + (rS - D)\,V_S - rV = 0. \tag{1}$$

(c) *Constraints*

We denote the general early pay-off for a path-dependent American option by $\Lambda(S, I, t)$. The absence of arbitrage opportunities implies that we must always have

$$V(S, I, t) \geqslant \Lambda(S, I, t).$$

If early exercise is possible, then (1) is only valid if early exercise is not desirable, that is when

$$V(S, I, t) > \Lambda(S, I, t).$$

If early exercise is desirable, this is so because it is more profitable to turn the option into its cash equivalent and put the funds in a bank; the implication is that $\mathscr{L}(V) < 0$ when early exercise is desirable. Thus,

$$\mathscr{L}(V) \leqslant 0;$$

the Black–Scholes differential operator can be seen to measure the difference between the riskless return on a hedged portfolio and the riskless return on a bank deposit; hence the equality when it is optimal to hold the option and the inequality when it is optimal to exercise it early. We may therefore cast the American option problem in the linear complementarity form

$$\left.\begin{aligned} V(S, I, t) \geqslant \Lambda(S, I, t), \quad \mathscr{L}(V) \leqslant 0, \\ (V(S, I, t) - \Lambda(S, I, t))\,\mathscr{L}(V) = 0. \end{aligned}\right\} \tag{2}$$

Uniqueness of the solution follows from continuity of V and V_S, prescription of the final value of V from the pay-off function $V(S,I,T)$ and boundary conditions that may apply as a result of barrier features. The linear complementarity form leads naturally to the formulation of the problem as a variational inequality, from which existence and uniqueness results may be easily deduced (see Wilmott *et al.* 1993).

(*d*) *Final conditions*

Equation (1) and the partial differential operator in (2) are backward parabolic and thus require final data to be specified at expiry, $t = T$. The appropriate final data is obviously the pay-off at expiry. Thus

$$V(S,I,T) = \Lambda(S,I,T).$$

The general form for Λ allows call, put and binary option varieties of exotics as well as of vanillas.

(*e*) *Boundary conditions*

For options with no barrier features, (2) must be solved in $0 \leqslant S < \infty$. In this case the transformation $S = S_0 e^x$, where S_0 is a typical value of S, maps $S = 0$ to $x = -\infty$, $S = \infty$ to $x = \infty$ and reduces the partial differential operator in (2) to a constant coefficient operator. This shows that we need not specify boundary conditions, other than a restriction on the growth as $x \to \pm\infty$. On the other hand, options with barrier features have boundaries at a finite value of S. In this case boundary conditions must be imposed. An example is a down-and-out barrier option, which we must solve on $X \leqslant S < \infty$, with $V = 0$ on $S = X$, the barrier.

(*f*) *Jump conditions*

When the average (for Asians) or the maximum and/or minimum (for lookbacks) is measured discretely, the governing partial differential equation contains delta functions in time. Except at sampling dates these vanish and the partial differential operator reduces to the usual Black–Scholes operator. Across sampling dates, t_i, the $\delta(t-t_i)\,V_I$ term can only be balanced by the V_t term, and hence across sampling dates we find that the option value locally satisfies the first order hyperbolic equation

$$V_t + f(S)\,\delta(t-t_i)\,V_I = 0. \tag{3}$$

This shows that the option value jumps across sampling dates and that a jump condition must be applied. These can be arrived at easily from equation (1) or alternatively by a simple financial argument. This argument is that the realized option price must be continuous across sampling dates. For example, for the discretely sampled geometric average option we have

$$V(S,I,t_i^-) = V(S,I+\log S,t_i^+).$$

This relates the value of the option before and after the sampling date t_i.

Similarly, if dividends are paid discretely, we can model this by a delta function in the dividend structure, $D(S,I,t) = \Sigma\,\delta(t-t_i)\,D_i(S,I)$. Across dividend dates the partial differential operator is approximated by

$$V_t - \delta(t-t_i)\,D(S,I)\,V_S = 0,$$

which again implies jump conditions. If the dividend structure is given, i.e. if $D(S,I)$ is given, then we can solve this equation across dividend dates to obtain the jump condition explicitly.

(*g*) *Further comments on pricing exotics*

More details about the partial differential equation approach to pricing exotic options can be found in Ingersoll (1987), Dewynne & Wilmott (1991, 1993) and Wilmott *et al.* (1993). There is one main advantage to this approach over the popular combination of equivalent martingale measure analysis and Monte Carlo simulation. This advantage is simply speed of computation. Once the problem has been formulated as a partial differential equation then we can apply quick numerical methods to calculate the price of an option as well as possibly to determine bounds on the error due to the discretization. Although the problems presented above are in three variables S, I and t, there are many important options for which there exist similarity reductions. This means that the numerical solution will generally be as quick as that for vanilla options; we shall see more of this later.

3. Transaction costs

(*a*) *Approaches to modelling the effects of transaction costs*

In the original Black–Scholes analysis and in that above for exotics, there are the assumptions of continuous rehedging and absence of transaction costs in trade in the underlying. Both of these assumptions are invalid in general and, depending on the liquidity of the market in question, may actually be very important. Several authors model option prices without these assumptions. In these circumstances we are faced with two distinct problems: determining the hedging strategy (when to rehedge) and valuing the option (given the hedging strategy). Rehedging will reduce, but not eliminate, risk, but at a cost. Whether we rehedge or not we must put a value to an inherently risky portfolio; we can no longer appeal to 'no-arbitrage'.

Two approaches have been taken in the academic literature: local in time and global in time. The former, for example Leland (1985), Boyle & Vorst (1992), Hoggard *et al.* (1993) and Whalley & Wilmott (1992), consider risk and return over a short interval of time. The latter, for example Hodges & Neuberger (1993) and Davis *et al.* (1993), adopt 'optimal strategies', in which risk and return are considered over the lifetime of the option. Both of these approaches have their advantages. The first group are easier to compute as they are only two-dimensional problems and, in particular, model market practice (Whalley & Wilmott 1992). The optimality of Hodges & Neuberger and Davis *et al.* is of obvious appeal, but leads to models that are usually impractically slow to compute. They also require input of the user's 'utility function', and very few practitioners adopt this approach. Recent unpublished research by Whalley and Wilmott suggests that simple asymptotic analysis of the global models, assuming that costs are in some sense small, yields nonlinear partial differential equations of the form encountered in the local models. Thus the two approaches may not be so different. For ease of exposition we shall describe the model of Leland and Hoggard *et al.* and then simply quote results from the other models.

(For an examination of a stochastic control problem in portfolio management with transaction costs see Morton & Pliska (1993), and for the asymptotic solution of that problem for small transaction costs see Atkinson & Wilmott (1993).)

(*b*) *The generalized Leland model*

We shall set up the model in a discrete time framework and assume that the asset price follows the random walk given by

$$\delta S = \mu S \, \delta t + \sigma S \phi \, \delta t^{\frac{1}{2}}, \tag{4}$$

where ϕ is a random number drawn from the standardized normal distribution. We are using $\delta \cdot$ to denote the small, but finite, change in a quantity. We can still write the option price as $V(S,t)$ and set up a portfolio as before with

$$\Pi = V - \Delta S,$$

with Δ to be chosen. We shall further assume that costs take the form

$$k_1 + k_2 N + k_3 NS,$$

where $N > 0$ is the number of shares traded to rehedge. Thus there are three components to the transaction costs: a fixed component (k_1), a cost proportional to the number traded ($k_2 N$) and a cost proportional to the value traded ($k_3 NS$).

Now consider the change in Π over the discrete time step δt. By expanding in Taylor series we can write

$$\begin{aligned}\delta \Pi &= V_t \, \delta t + V_S \, \delta S + \tfrac{1}{2} V_{SS} \, \delta S^2 - \Delta \, \delta S - (k_1 + k_2 N + k_3 NS) + \ldots \\ &= V_t \, \delta t + (V_S - \Delta) \, \delta S + \tfrac{1}{2} \sigma^2 S^2 \phi^2 \, \delta t V_{SS} - (k_1 + k_2 N + k_3 NS) + \ldots.\end{aligned}$$

Note the similarity to the usual Black–Scholes argument. There are important differences however. First, we have subtracted off the transaction costs; they cause a reduction in the value of the portfolio. Second, the V_{SS} term is multiplied by the square of the random variable ϕ; we cannot appeal to Itô's lemma in this discrete time world.

Now we come to choosing the hedging strategy. The Leland strategy is to take

$$\Delta = V_S, \tag{5}$$

as in Black–Scholes, and rehedge every time step. (The choice (5) minimizes the variance, and hence the risk, of the portfolio to the order of magnitude we are considering.)

We now have

$$\delta \Pi = V_t \, \delta t + \tfrac{1}{2} \sigma^2 S^2 \phi^2 \, \delta t V_{SS} - (k_1 + k_2 N + k_3 NS) +$$

The change in the portfolio value, $\delta \Pi$, still contains elements of risk, in the ϕ^2 term and the transaction cost term. The latter is because we do not know N. For this reason we work in terms of the expected change in the value of the portfolio. Thus

$$E[\delta \Pi] = V_t \, \delta t + \tfrac{1}{2} \sigma^2 S^2 \, \delta t V_{SS} - (k_1 + k_2 E[N] + k_3 E[N] S) + \ldots. \tag{6}$$

The only unknown term in (6) is $E[N]$, and this is easily calculated. Because $\Delta = V_S(S,t)$, we know that

$$N = |\Delta(S + \delta S, t + \delta t) - \Delta(S,t)|;$$

this is the change in the number of shares held. Thus

$$N = |V_S(S + \delta S, t + \delta t) - V_S(S,t)|,$$

and, to leading order, $$N = |V_{SS}\,\sigma S\phi\,\delta t^{\frac{1}{2}}|.$$

From this it follows that

$$E[N] = \surd(2/\pi)\,\sigma S\,\delta t^{\frac{1}{2}}|V_{SS}|.$$

We now value the option by setting the return on the portfolio equal to that from a risk-free deposit. This is our valuation policy; we have eliminated risk to the best of our ability given our hedging strategy and we attach no further value to accepting the remaining risk. We thus find that

$$V_t + \tfrac{1}{2}\sigma^2 S^2 V_{SS} + rSV_S - rV = (k_1/\delta t) + (k_2 + k_3 S)\surd(2/\pi\delta t)\,\sigma S|V_{SS}|, \tag{7}$$

where we have dropped terms of $O(\delta t^{\frac{1}{2}})$ or smaller.

(c) *Other models and matters arising*

The above model is an extended version of that due to Leland. The correction due to the effect of transaction costs depends on the second derivative of the option price with respect to the asset price. This derivative is commonly referred to as the 'gamma', and it is a measure of the mishedging due to the discreteness of the hedging, and hence a measure of the level of transaction costs.

Because of the significance of V_{SS}, many transaction cost models result in equations of the form

$$V_t + \tfrac{1}{2}\sigma^2 S^2 V_{SS} + rSV_S - rV = F(S, V_{SS}),$$

where the function $F(\cdot,\cdot)$ depends upon the hedging strategy. We shall now give some examples.

(i) Extended Leland (see Hoggard *et al.* 1993): rehedge to Black–Scholes delta every δt,

$$F(S,x) = (k_1/\delta t) + (k_2 + k_3 S)\surd(2/\pi\delta t)\,\sigma S|x|.$$

(ii) Market practice model (see Whalley & Wilmott 1992): continuous time, rehedge to Black–Scholes value when delta moves outside a 'hedging bandwidth' of size ϵ/S, where ϵ is a function of S, t as well as V and its derivatives,

$$F(S,x) = (\sigma^2 S^4/\epsilon)\,[k_1 + (k_2 + k_3 S)\,\epsilon^{\frac{1}{2}}/S]\,x^2.$$

A special case of this is $\epsilon = \sigma^2 S^4 V_{SS}^2\,\delta t^{\frac{1}{2}}$, in which case the expected time between rehedges is constant and the model reduces to the extended Leland.

(iii) Small costs limit of Davis *et al.* (1993) model: this model is in continuous time and is based upon utility maximization,

$$F(S,x) = \frac{e^{-r(T-t)}}{\gamma}\left(\frac{3k_3\gamma^2 S^4\sigma^3}{8\delta^2}\right)^{\frac{2}{3}}\left(\left|x - \frac{e^{-r(T-t)}(\mu - r)}{\gamma S^2\sigma^2}\right|\right)^{\frac{4}{3}}.$$

Here γ is the index of risk aversion and $\delta = e^{-r(T-t)}$. Note that this problem depends on the growth rate μ.

In Whalley & Wilmott (1992) many issues arising from such equations are discussed. Briefly, these include the following.

1. Nonlinearity. As the right-hand side of the equation is in each case a nonlinear function of the Black–Scholes value of gamma, V_{SS}, there will inevitably be different

values for short and long positions. Also portfolios of options should be treated as a whole and not as the sum of individually valued components. There will be advantage to be gained from offsetting opposite positions as well as economies of scale.

2. Negative option prices. With the general cost structure discussed in Hoggard *et al.* (1993) and Whalley & Wilmott (1992) (not simply bid-offer spread) it is possible to arrive at negative option prices. (To see this, consider the commission component of costs. If a fixed amount is paid at each rehedge then for small asset values it may cost more to hedge a call than the call is worth.) This suggests modifying hedging strategies to allow the possibility of not rehedging if to rehedge would make the option value negative. This introduces a free boundary below which (for a call) the option should not be rehedged.

3. American options. As also mentioned in Davis *et al.* (1993), it is the owner of the American option who controls its exercise. It is difficult to value an American option optimally unless the owner's hedging and exercise strategy is known. This entails at least knowing all of the owner's estimates of the parameters.

The strong nonlinearities associated with the transaction cost term $F(S, V_{SS})$ can lead to ill-posedness. We interpret this as meaning that the hedging strategy is inappropriate. As an example, we consider the market practice model (Whalley & Wilmott 1992) with $k_1 \ll 1$ and $k_2 = k_3 = 0$. Thus we have

$$V_t + \tfrac{1}{2}\sigma^2 S^2 V_{SS} + rSV_S - rV = (k_1 \sigma^2 S^4/\epsilon)(V_{SS})^2.$$

We assume that $\delta = k_1 \sigma^2/\epsilon \ll 1$ so that transaction costs appear as a small perturbation to the Black–Scholes equation. We see, however, that the sign of the curvature, that is, the sign of the gamma, is critically important. Solving for V_{SS} we find that

$$S^2 V_{SS} = \tfrac{1}{2}\sigma^2/\delta[1 \pm \surd(1 + (4\delta/\sigma^4)(V_t + rSV_S - rV))].$$

On the one hand, if $V_{SS} < 0$ then

$$S^2 V_{SS} \sim -(1/\sigma^2)(V_t + rSV_S - rV) + O(\delta),$$

and we have a backward parabolic problem with final conditions. This is well posed. Moreover, in this situation we are justified in looking for a regular asymptotic expansion,

$$V = V_0 + \delta V_1 + O(\delta^2),$$

which treats the hedging cost as a small perturbation to the Black–Scholes model. If the gamma is negative we can use this model to value the option.

On the other hand, if $V_{SS} > 0$ we have

$$S^2 V_{SS} \sim (\sigma^2.\delta) + (1/\sigma^2)(V_t + rSV_S - rV) + O(\delta),$$

which then gives us a forward parabolic problem with final conditions. This is ill posed. Moreover, we see that we must look for asymptotic solutions of the form

$$V = (V_{-1}/\delta) + V_0 + I(\delta),$$

whenever V_{SS} is $O(\delta^{-1})$, as in the case of an at the money call or put near expiry. Thus there is a blow up (associated with the ill-posedness) which will, in most cases, occur over a very short time scale in the neighbourhood of the expiry date. This shows that we cannot value an option hedged according to the market practice model if the gamma is large and positive.

4. Valuing exotic options in the presence of transaction costs

The modelling and analyses which led to the exotic option problem and the transaction cost problem separately may be formally combined to give the following nonlinear partial differential equation problem for exotic options in the presence of transaction costs.

$$V_t + fV_I + \tfrac{1}{2}\sigma^2 S^2 V_{SS} + (rS - D)\, V_S - rV = F(S, V_{SS}), \tag{8}$$

where we can take F to be any of the functions in the previous section depending on the hedging strategy.

(a) *Similarity solutions*

In general the solution for the value of the path-dependent options we have mentioned must be solved numerically and usually in the three dimensions of asset price, path-dependent quantity and time. In some special cases however there are similarity reductions that reduce the number of dimensions to only two. This significantly decreases the computing time. Whether there is a similarity reduction not only depends on the pay-off but also on the nature of the hedging strategy and the type of costs involved. This is easily demonstrated by example.

It is well known and easily seen that in the absence of transaction costs the lookback put having pay-off

$$\max\,(J - S, 0),$$

where J is the maximum, has a similarity solution of the form

$$V(S, J, t) = JH(S/J, t).$$

There is still a similarity solution of this form in the Leland hedging model with only the bid-offer spread cost. If either of the k_1 and k_2 cost terms are non-zero then this solution is rendered invalid.

Another example is the average strike foreign exchange option having pay-off

$$\max\,[1 - (I/TS), 0],$$

where

$$I = \int_0^t S\, \mathrm{d}\tau.$$

This has a similarity solution of the form

$$V(S, I, t) = H(S/I, t)$$

in the absence of transaction costs and the same form in the Leland model when $k_2 = 0$.

Despite these examples being faster to compute they require special restrictions on the cost structure. For this reason the examples we shall shortly be giving are all genuinely three-dimensional.

(b) *Discrete sampling and discrete dividends*

We have set up an exotic pricing model that allows for discretely paid dividends and discretely sampled path-dependent quantities. We have then introduced transaction costs and described the small effects of transaction costs. Are these consistent? Let us consider the case of discretely paid dividends; the following idea carries over directly to the discretely sampled exotic case.

Recall from earlier that across a dividend state the realized option price is continuous but V as a function of S jumps. This is because the realized path of the asset is itself discontinuous across a dividend date. Even though the realized option price is continuous, the delta of the option (its derivative with respect to the underlying asset) is discontinuous. Thus across a dividend date there must be a rehedge of the order of the dividend yield. Each time there is a dividend payment there is a rehedge of this order. Typically, therefore, the total cost associated with these relatively large rehedges is, in the Leland model for example, of order

$$Mk_1 + (k_2/S) \times \text{total dividend} + k_3 \times \text{total dividend},$$

where M is the number of dividend payments. On the other hand, we have already seen that the rehedging over each period δt leads to the total costs over the life of the option of order

$$(k_1 T/\delta t) + [k_2 T\sigma/\surd(\delta t)] + [k_3 ST\sigma/\surd(\delta t)].$$

From these it can be seen that in practice, the total costs associated with a small number of discrete dividend payments are small compared with the accumulation of costs due to the frequent rehedging between dividend dates. We shall thus ignore them.

5. Results

In this section we consider numerically computed solutions of the generalized Leland model for lookback strike puts. The maxima, J, for the underlying are assumed to be discretely sampled, that is

$$J = \max_i \{S(t_i)\},$$

where t_i are the sampling dates. The solutions have been calculated using explicit finite differences, which are necessary in view of the strong nonlinearities introduced by the transaction costs. The jump condition across a maximum sampling date is

$$V(S, J, t^-) = V(S, \max(J, S), t^+)$$

(see, for example, Dewynne & Wilmott 1991). Note that this implies we must solve the problem on a square grid in the (S, J) plane. An implication is that we cannot use asymptotic estimates for $V(S, J, t)$, which apply only for $S \gg J$, when S and J are comparable. We overcome this difficulty by using explicit finite differences and not applying boundary conditions for large S. Rather, at each time step we discarded the value of V for the largest values of S and J, that is, our square grid shrank by one mesh step at each time step.

In figures 1, 2 and 3 we value long positions in discretely sampled lookback strike puts. The pay-offs are

$$\max(J - S, 0).$$

The lifetimes of the options are all $T = 0.25$, the annual volatility is 0.2, the annual interest rate is 0.05 and there are eight discrete times at which the maximum is sampled. These are at $t = 0.225$, $t = 0.200$, and so on back to $t = 0.025$. In all cases we assume the option value may not become negative.

In figure 1 there are no transaction costs, $k_1 = k_2 = k_3 = 0$. We plot the option

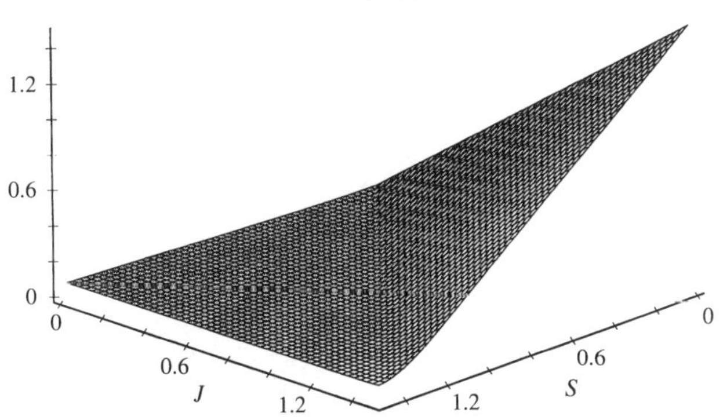

Figure 1. The value of a discretely sampled lookback strike put with pay-off max $(J-S, 0)$ as a function of J and S.

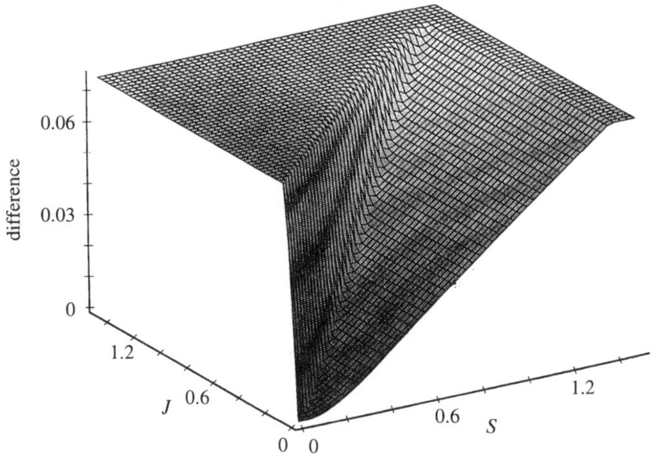

Figure 2. The difference between the Black–Scholes and with-cost value of a discretely sampled lookback strike put with payoff max $(J-S, 0)$ as a function of J and S.

value against S and J. This problem has a similarity solution with independent variable S/J. If V/J is plotted against this variable then the picture would show a minimum close to $S/J = 1$. The delta for this option thus changes sign. This is because the holder of the option benefits from a large value of the maximum J established before expiry but a small value of S at expiry, and, of course, these two are not independent.

In figure 2 there are fixed transaction costs only; $k_1/\delta t = 0.3$, but $k_2 = k_3 = 0$. We plot the difference between the Black–Scholes (zero cost) value and the solution with costs. The difference is always positive; the option's value has been decreased. The effect of the fixed transaction cost is to decrease the option value everywhere by the same amount. This gives negative option values for certain S and J. However, as soon as the positivity constraint is added the option value is no longer decreased everywhere by the same amount; it is decreased by less at some places. This explains the dip in the plot. This problem does not have a similarity solution. As discussed

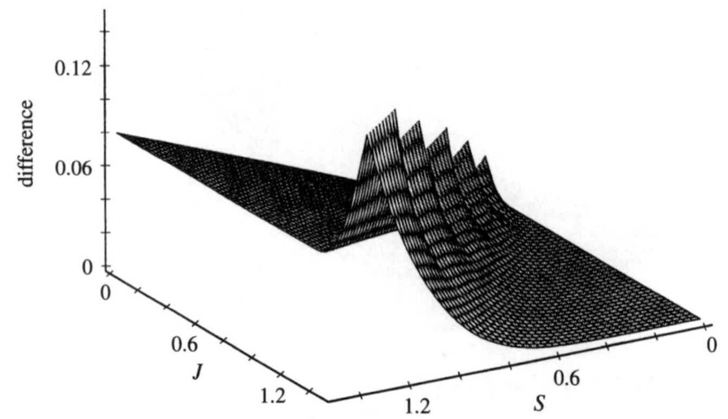

Figure 3. The difference between the Black–Scholes and with-cost value of a discretely sampled lookback strike put with payoff $\max(J-S,0)$ as a function of J and S.

above the option price would naturally become negative for some values of S and J. Here we have imposed the constraint $V \geqslant 0$. For some values of S and J the option price is zero; we have solved a free boundary problem.

In figure 3 there are only costs proportional to the volume of underlying traded, $k_1 = 0$, $k_2 \sigma \surd(2/\pi\delta t) = 0.3$ and $k_3 = 0$. Again we have plotted the difference between the theoretical Black–Scholes value and the value allowing for costs. The effect of costs is again to decrease the value of the option. For small S the cost effect tends to zero, since it is proportional to the option's gamma. For large S the option value is zero, and so the difference is simply the large S behaviour of the Black–Scholes solution. Again there is no similarity solution. Since the second derivative V_{SS} is positive the diffusion term vanishes at

$$\tfrac{1}{2}\sigma^2 S^2 = k_2 S\sigma \surd(2/\pi\delta t),$$

that is

$$S = (2k_2/\sigma)\surd(2/\pi\delta t).$$

We thank the Royal Society and Adam, Harding & Lueck for their support.

References

Atkinson, C. & Wilmott, P. 1993 Portfolio management with transaction costs: an asymptotic analysis. *Math. Finance*. (In the press.)

Black, F. & Scholes, M. 1973 The pricing of options and corporate liabilities. *J. Political Econ.* **81**, 637–654.

Boyle, P. P. & Vorst, T. 1992 Option replication in discrete time with transaction costs. *J. Finance* **47**, 271.

Davis, M. H. A., Panas, V. G. & Zariphopoulou, T. 1993 European option pricing with transaction costs. *SIAM Jl. Control* **31**, 470–493.

Dewynne, J. N. & Wilmott, P. 1991 Lookback options. OCIAM, Oxford (Working paper).

Dewynne, J. N. & Wilmott, P. 1993 Partial to the exotic. *Risk* March, 38–46.

Hodges, S. D. & Neuberger, A. 1993 Optimal replication of contingent claims under transaction costs. *Rev. Future Markets* **8**, 222–239.

Hoggard, T., Whalley, A. E. & Wilmott, P. 1993 Hedging option portfolios in the presence of transaction costs. *Adv. Fut. Opt. Res.* (In the press.)

Ingersoll, J. 1987 *Theory of financial decision making*. Rowman and Littlefield.

Leland, H. E. 1985 Option pricing and replication with transaction costs. *J. Finance* **40**, 1283–1301.

Morton, A. & Pliska, S. R. 1993 Optimal portfolio management with fixed transaction costs. *Math. Finance*. (In the press.)

Whalley, A. E. & Wilmott, P. 1992 A hedging strategy and option valuation model with transaction costs. OCIAM, Oxford (Working paper).

Wilmott, P., Dewynne, J. N. & Howison, S. D. 1993 *Option pricing: mathematical models and computation*. Oxford Financial Press.

8

Stochastic equity volatility and the capital structure of the firm

A. Bensoussan, M. Crouhy and D. Galai

This paper develops a general model for equity volatility when the firm is financed by equity, debt and any other financial instruments like warrants and convertible bonds. The stochastic nature of equity volatility is endogenous and comes from the impact of a change in the value of the firm's assets on the financial leverage. We first present the basic model to value corporate securities, which is an extension of the Black–Scholes model. Then, we are able to propose an analytic approximation for equity volatility, which is shown to be extremely precise. Finally, we study the behaviour of equity volatility when the firm is financed by equity and debt.

1. Introduction

Although the Black–Scholes (1973) model (BSM) is very popular to price equity options and other complex derivative securities, its normative and theoretical requirements are very restrictive. One of the major assumptions needed to derive the BSM is that the value of the underlying asset follows a stationary geometric brownian motion with constant variance. However, for any non-pure equity firm the variance of equity quite probably will not be stationary. Because the volatility of equity is non-stationary, it cannot be used as the underlying asset in BSM for pricing contingent liabilities of the firm or any other derivative. The study of the behaviour of the volatility function for equity will lead to better approximation methods.

In this paper we study the behaviour of equity volatility when its stochastic nature is endogenous and stems from the impact of a change in the value of the firm's assets on the financial leverage. In the next section we present the basic model to value corporate securities. Then, in the third section, we are able to propose an analytic approximation for equity volatility which is shown to be extremely precise. Finally, in the fourth section we look at the behaviour of equity volatility, when the firm is financed by equity and debt.

2. The basic model

Consider a firm whose total asset value $V(t)$ is solution of the Black–Scholes risk-neutral stochastic differential equation,

$$dV/V = r\,dt + \bar{\sigma}\,dz, \tag{1}$$

where r is the continuously compounded risk-free rate of interest which is assumed to be a constant, $\bar{\sigma}^2$ is the instantaneous variance of the return on the asset, which is also assumed to be a constant, t is the current time and z is a standard Wiener

Mathematical Models in Finance. Edited by S.D. Howison, F.P. Kelly and P. Wilmott.
Published in 1995 by Chapman & Hall, London. ISBN 0 412 63070 2

process. $V(t)$ according to equation (1), is assumed to follow a stationary log-normal distribution with a constant variance such that the discounted price is a martingale. The focus in this paper is on the behaviour of equity volatility when the firm is financed by debt and equity. The model can be extended to more complex capital structures which, for example, include warrants like in Crouhy & Galai (1994) and Bensoussan *et al.* (1993). $V(t)$ is the only exogenous source of risk, but it is generally unobservable and non-traded. Indeed, several components of the firm's capital structure, such as bank loans are not traded on an exchange.

Whenever the assets are financed by equity and additional sources of funds like debt, the stochastic processes for the value of the assets $V(t)$, and of equity $S(t)$, do not coincide. However, developing pricing models in tem of $S(t)$ as the underlying asset does not allow the use of BSM anymore because the stochastic process for $S(t)$ violates the constant volatility assumption.

Our basic hypothesis is that there exists a deterministic functional relation between $S(t)$ and $V(t)$, namely

$$S(t) = S(V(t), t), \tag{2}$$

that implies that $S(t)$ follows a markovian process. Moreover, we assume that there is no arbitrage opportunity so that the function $S(V, t)$ satisfies the traditional Black–Scholes partial differential equation (PDE),

$$-S_t - \tfrac{1}{2}\bar{\sigma}^2 V^2 S_{VV} - rVS_V + rS = 0. \tag{3}$$

To complete the characterization of $S(V, t)$, for $t < T$, it is sufficient to specify the value of $S(V, T)$ at time T. At this stage it is left as a parameter with the condition $S(V, T) \geqslant 0$, so that for any t, $t < T$, $S(V, t) \geqslant 0$ and $\not\equiv 0$.

The only restriction we impose is that we can inverse the function $S(V, t)$, which means that equation $S = S(V, t)$ has a unique solution $V = V(S, t)$. Actually in a levered firm, equity is a monotonic convex function of V, and as a consequence this restriction is satisfied. (In what follows the same notation S or V will be used for an independent argument as well as for the function $S(V, t)$ or $V(S, t)$. Although this may be a source of confusion this preserves the mnemonics.) From the implicit function theorem, we know that the local existence and uniqueness of $V(S, t)$ at any point (V_0, S_0) such that $S_0 = S(V_0, t)$ is guaranteed, provided the derivative $S_V(V_0, t)$ is not equal to zero. In practice, this is not a very stringent assumption since we expect the value of equity to grow with the total asset value. In fact, from (3) the possibility that $S_V(V, t) = 0$ can occur only at isolated points, unless $S(V, t)$ does not depend at all on V. Then, the following identity holds:

$$S \equiv S(V(S, t), t) \quad \text{for all } S \text{ in the domain of } V(S, t), \tag{4}$$

and as a consequence, we also obtain

$$V \equiv V(S(V, t), t), \quad \text{for all } V \text{ in the range of } V(S, t). \tag{5}$$

By differentiating (4) we get

$$V_S(S, t) = 1/S_V(V(S, t), t).$$

From the assumptions concerning the stochastic process of $V(t)$ and using Itô's calculus, it can be shown that the return on equity follows a diffusion described by the following stochastic differential equation

$$\mathrm{d}S/S = r\,\mathrm{d}t + \sigma\,\mathrm{d}z, \tag{6}$$

where σ denotes the instantaneous volatility of the return on equity, such that

$$\sigma = \sigma(S,t) = \bar{\sigma} S_V(V(S,t),t)\frac{V(S,t)}{S} = \frac{\bar{\sigma}}{S}\frac{V(S,t)}{V_S(S,t)}. \tag{7}$$

Therefore, the value of equity $S(t)$ follows a model similar to $V(t)$ except for its volatility which is no longer a constant, and is precisely given by the function $\sigma(S,t)$ as specified in (7) where the argument S has to be replaced for the value of the process $S(t)$. (For a proof, see Bensoussan *et al.* (1993).)

It is also useful to introduce $\tilde{\sigma}$, the volatility of S as a function of V, while σ denotes the volatility of S as a function of S.

$$\tilde{\sigma} = \tilde{\sigma}(V,t) = \bar{\sigma} S_V(V,t)\, V/S(V,t). \tag{8}$$

We then have the correspondence

$$\begin{cases} \sigma(S,t) = \tilde{\sigma}(V(S,t),t), \\ \tilde{\sigma}(V,t) = \sigma(S(V,t),t). \end{cases} \tag{9}$$

This property will be useful in the next section to derive an analytic approximation for equity volatility as a function of the stock price.

We observe that when V is the underlying variable, $\bar{\sigma}$ the volatility of the return process for V appears in the no-arbitrage equilibrium condition (3) for any asset price $S(V,t)$. If instead we consider S as the underlying variable, the condition for no arbitrage equilibrium for the value function $V(S,t)$ is, as the next lemma shows, the solution of a Black–Scholes type PDE similar to (3), except the volatility term is $\sigma = \sigma(S,t)$, as defined by (7).

Lemma 2.1. *The function $V(S,t)$ defined as the inverse of $S(V,t)$ is solution of the* PDE

$$-V_t - \tfrac{1}{2}\sigma^2 S^2 V_{SS} - rSV_S + rV = 0. \tag{10}$$

Proof. By differentiating equation (4) twice with respect to S, we obtain

$$1 = S_V V_S \quad \text{so that} \quad S_V = 1/V_S,$$
$$0 = V_{SS} S_V + (V_S)^2 S_{VV}, \quad \text{which gives} \quad S_{VV} = -V_{SS}/V_S^3.$$

Then, differentiating (4) with respect to time, it follows that $S_t = -S_V V_t$.

Replacing in equation (3) S_t, S_V and S_{VV} by the above terms and rearranging them, we obtain (10). □

3. Approximating equity volatility

In this section we first show that the functions $\sigma(S,t)$ and $\tilde{\sigma}(V,t)$ are solutions of nonlinear PDEs for which there is no closed-form solution and which are quite complex to solve numerically. However, we are able to derive analytic approximations which are shown to be very accurate. (To the knowledge of the authors, this is the first time that PDEs for the volatility measures (11) and (12), appear in the option pricing literature.)

Proposition 3.1. *$\sigma(S,t)$ satisfies the* PDE,

$$-\sigma_t - \tfrac{1}{2}\sigma^2 S^2 \sigma_{SS} - \sigma_S S(r + \sigma^2) = 0, \tag{11}$$

and $\tilde{\sigma}(V t)$ is the solution of the PDE,

$$-\tilde{\sigma}_t - \tfrac{1}{2}\bar{\sigma}^2 V^2 \tilde{\sigma}_{VV} - \tilde{\sigma}_V V(r + \bar{\sigma}\tilde{\sigma}) = 0. \tag{12}$$

Of course, the boundary values for $\sigma(S, T)$ and $\tilde{\sigma}(V, T)$ have to be specified in order to solve (11) and (12). They follow from (7) and (8) when $V(S, T)$ and $S(V, T)$ are made explicit. The proof is given in Appendix A.

(a) Approximation of the volatility function

Equation (11), which defines equity volatility $\sigma(S, t)$, is strongly nonlinear. Not only does it not have any analytic solution, but it is difficult to solve numerically. In this section we propose a very accurate analytical approximation for $\sigma(S, t)$.

We first introduce the functions:

$$\theta(S, t) = S.(\sigma(S, t) - \bar{\sigma}), \quad \tilde{\theta}(V, t) = S(V, t).(\tilde{\sigma}(V, t) - \bar{\sigma}). \tag{13}$$

We next show that θ and $\tilde{\theta}$ are solutions of Black–Scholes PDEs similar to (10) and (3), respectively, although the boundary conditions are obviously different.

Proposition 3.2.

$$-\theta_t - \tfrac{1}{2}\sigma^2 S^2 \theta_{SS} - \theta_S Sr + r\theta = 0, \tag{14}$$

$$-\tilde{\theta}_t - \tfrac{1}{2}\bar{\sigma}^2 V^2 \tilde{\theta}_{VV} - \tilde{\theta}_V Vr + r\tilde{\theta} = 0. \tag{15}$$

The proof is straightforward.

To simplify (14) further so that we keep only one unknown, θ, instead of two, θ and σ, we replace in (14) σ by its definition (13) in term of θ, i.e.

$$\sigma = \bar{\sigma} + (\theta/S). \tag{16}$$

Now, consider the function $\tilde{\theta}(S, t)$ which should not be confused with $\theta(S, t)$, as by the correspondence law (9), we have:

$$\theta(S, t) = \tilde{\theta}(V(S, t), t) \neq \tilde{\theta}(S, t),$$

$$\tilde{\theta}(S, t) = \theta(S(S, t), t) \neq \theta(S, t).$$

If in (15) we replace the argument V by S, then $\tilde{\theta}(S, t)$ is solution of the equation

$$-\tilde{\theta}_t - \tfrac{1}{2}\bar{\sigma}^2 S^2 \tilde{\theta}_{SS} - \tilde{\theta}_S Sr + r\tilde{\theta} = 0. \tag{17}$$

But then (17) can be viewed as (14) after σ has been replaced according to (16) and the term in θ/S has been deleted. This argument provides a first intuition of why $\tilde{\theta}(S, t)$ (solution of PDE 17) might be a good approximation of $\theta(S, t)$. We formally justify this approximation in the next subsection. However, to solve (17) we need to specify its boundary condition at time T, which from (9) and (13) can be written as

$$\tilde{\theta}(S, T) = \theta(S(S, T), T).$$

However, it is generally the case that for any value of S, $S(S, T) \neq S$, so that $\theta(S, T) \neq \theta(S(S, T), T)$. If it is more convenient to compute $\theta(S, T)$, it may be tempting to take as an approximation of $\tilde{\theta}(S, t)$ the solution of (17) with the boundary condition $\theta(S, T)$ instead of $\theta(S, (S, T), T)$. It may happen (this is precisely the case of equity volatility in a firm financed by equity and debt) that $\theta(S, T)$ and $\theta(S(S, T), T)$ coincide although $S \neq S(S, T)$ for all S.

If we take $\tilde{\theta}(S,t)$ as an approximation of $\theta(S,t)$, from (13) the corresponding approximation for $\sigma(S,t)$ is

$$\sigma^*(S,t) = \bar{\sigma} + [\tilde{\theta}(S,t)/S]. \tag{18}$$

This is not $\tilde{\sigma}(S,t)$. In fact, from (13) we obtain

$$\sigma^*(S,t) = \bar{\sigma} + S(S,t)(\tilde{\sigma}(S,t) - \bar{\sigma})/S. \tag{19}$$

Note that $\theta(S,t)$ and $\tilde{\theta}(S,t)$ are not small in general. However, the term $\theta(S,t)/S$ is small for large values of S, in particular when $\theta(S,t)$ is bounded so $\bar{\sigma}$ is already an approximation of $\sigma(S,t)$, but $\sigma^*(S,t)$ is shown to be a much better approximation which is, for practical purposes, extremely accurate.

Indeed, the approximation error for equity volatility is

$$\sigma^*(S,t) - \sigma(S,t) = [\tilde{\theta}(S,t) - \theta(S,t)]/S, \tag{20}$$

and therefore, if the approximation of $\theta(S,t)$ by $\tilde{\theta}(S,t)$ is accurate, the approximation of $\sigma(S,t)$ by $\sigma^*(S,t)$ is even better by the factor $1/S$. On the other hand, the error when we approximate $\sigma(S,t)$ by $\bar{\sigma}$ is $\theta(S,t)/S$, which is much larger than (20).

(*b*) *Justification of the approximation of* $\theta(S,t)$ *by* $\tilde{\theta}(S,t)$

The approximation error is defined by the difference between the approximation and the exact value, i.e.

$$e(S,t) = \tilde{\theta}(S,t) - \theta(S,t)$$

and by the correspondence law (9), we can write $e(S,t)$ as

$$e(S,t) = \tilde{\theta}(S,t) - \tilde{\theta}(V(S,t),t). \tag{21}$$

Theorem 3.1 shows the order of magnitude of the approximation error. To prove this theorem intermediate results are needed and they are shown in Appendix B.

Theorem 3.1. *If there exists a constant* a, *such that for any constant* M,

$$\left.\begin{array}{l} V(S_V(V,T)-a) \to 0 \quad as \quad V \to \infty, \\ |V(S_V(V,T)-a)| \leqslant M, \end{array}\right\} \tag{22}$$

and if also

$$[\max(V(S,t),S)]/\min(V(S,t),S) \leqslant k \, \forall \, S \geqslant S_0$$

then the error term $e(S,t)$ *has the property*

$$e(S,t) \to 0 \quad as \quad S \to \infty.$$

Proof. From expression (21), for the approximation error we can write

$$e(S,t) = (S - V(S,t)) \int_0^1 \tilde{\theta}[V(S) + \lambda(S - V(S)), t] \, \mathrm{d}\lambda. \tag{23}$$

Consider a sequence S^n which tends to $+\infty$ as $n \to \infty$. We note from proposition B.2 that for any λ,

$$\tilde{\theta}_V[V(S^n) + \lambda(S^n - V(S^n)), t] \times [V(S^n) + \lambda(S^n - V(S^n))] \to 0 \quad \text{as} \quad n \to \infty$$

and is bounded. Moreover, since we may assume that $S^n \geqslant S_0$, then

$$\left| \frac{[S^n - V(S^n)]}{[V(S^n) + \lambda(S^n - V(S^n))]} \right| \leqslant k.$$

We can then apply Lebesgue's theorem in the integral (23), giving the conclusion that

$$e(S^n, t) \to 0. \qquad \square$$

In fact the error tends to 0 very fast, as we shall illustrate in the next section.

4. Equity volatility for a firm financed with equity and debt

Consider a firm whose assets, V, are financed by equity, S, and zero-coupon debt, D. The face value of debt is F and it will mature at time T. Following Black & Scholes (1973), Merton (1973) and Galai & Masulis (1976), we can consider the value of equity as a European call option on the assets of the firm, with terminal value at maturity T of debt, $S(V,T) = (V-F)^+$.

Denote by $D(S,t)$ the value of debt at any time $t, t \leqslant T$. Hence, by the correspondence law (9),

$$\tilde{D}(V,t) = D(S(V,t),t).$$

At time T we have

$$\tilde{D}(V,T) = \min(V,F) = V - S(V,T)$$

and, of course, the following balance sheet identity holds.

$$V \equiv S(V,T) + \tilde{D}(V,T).$$

The value of stock, $S(V,t)$, is given by the Black–Scholes formula

$$S(V,t) = BS(V,t \mid F, T, \bar{\sigma}, r),$$

where (keeping notations simple by omitting the parameters F, T, $\bar{\sigma}$ and r on which BSM is conditional),

$$\left.\begin{aligned} BS(V,t) &= VN(d_1) - F\,\mathrm{e}^{-r\tau}N(d_2), \\ d_1 \equiv d_1(V) &= \frac{1}{\bar{\sigma}\sqrt{\tau}}\left(\log\left(\frac{V}{F}\right) + (r + \tfrac{1}{2}\bar{\sigma}^2)\,\tau\right), \\ d_2 \equiv d_2(V) &= \frac{1}{\bar{\sigma}\sqrt{\tau}}\left(\log\left(\frac{V}{F}\right) + (r - \tfrac{1}{2}\bar{\sigma}^2)\,\tau\right), \\ \tau &= T - t, \end{aligned}\right\} \tag{24}$$

and where $N(\,)$ denotes the cumulative standard normal distribution. The Black–Scholes formula is the solution of (3) given the boundary condition $S(V,T) = (V-F)^+$. The derivative of the Black–Scholes formula with respect to the underlying variable V is

$$BS_V(V,t) = N(d_1) \geqslant 0, \tag{25}$$

so that $S(V,t)$ is strictly monotonic in V for all $t < T$. Hence the function $V(S,t)$ is well defined for $t < T$. Note that at maturity T, if $V > F$, then $S > 0$,

$$V(S,T) = S + F, \quad \text{and} \quad S_V(V,T) = \mathbb{I}_{V \geqslant F}, \tag{26}$$

where $\mathbb{I}_{\text{condition}}$ is an indicator function which is equal to one when the condition is satisfied, and 0 otherwise. Using formula (41) in Appendix B we get

$$\tilde{\theta}_V(V,t) = \frac{1}{\sqrt{(2\pi\tau)}}\frac{1}{\tau\bar{\sigma}^2}\int_{-\bar{\sigma}\sqrt{(\tau)}d_1(V)}^{+\infty} x\exp\left(-\frac{1}{2}\frac{x^2}{\tau\bar{\sigma}^2}\right)\mathrm{d}x.$$

Hence,

$$\tilde{\theta}_V(V,t) = \frac{1}{\sqrt{(2\pi\tau)}} \exp[-\tfrac{1}{2}(d_1(V))^2],$$

and by using (24) it is also equal to

$$\tilde{\theta}_V(V,t) = \frac{F\mathrm{e}^{-r\tau}}{\sqrt{(2\pi\tau)}} \frac{1}{V} \exp[-\tfrac{1}{2}(d_2(V))^2]. \tag{27}$$

We then see that condition (44) in Appendix B holds, and as a consequence the approximation error $e(S,t)$ tends to zero as S becomes large. In fact, we can derive an explicit expression for the error so that we can assess precisely the error for any value of S. Note that we can write (27) as

$$\tilde{\theta}_V(V,t) = \bar{\sigma} F \mathrm{e}^{-r\tau} \frac{\mathrm{d}}{\mathrm{d}V}[N(d_2(V))].$$

Hence

$$\tilde{\theta}(S,t) = \bar{\sigma} F \mathrm{e}^{-r\tau} N(d_2(S)), \tag{28}$$

and from (18) the volatility functions $\sigma(S,t)$ will be approximated by

$$\sigma^*(S,t) = \bar{\sigma}\left(1 + \frac{F}{S}\mathrm{e}^{-r\tau} N(d_2(S))\right). \tag{29}$$

The approximation error is

$$e(S,t) = \tilde{\theta}(S,t) - \tilde{\theta}(V(S,t),t) = \bar{\sigma} F\, \mathrm{e}^{-r\tau}\{N(d_2(S)) - N[d_2(V(S,t))]\}. \tag{30}$$

To assess the magnitude of $e(S,t)$ we need bounds on $V(S,t)$. From Black–Scholes' boundary conditions, we have

$$S \leqslant V(S,t) \leqslant S + F\, \mathrm{e}^{-r\tau}. \tag{31}$$

Hence, from (30) we can derive the bounds for the approximation error

$$0 \geqslant e(S,t) \geqslant -\bar{\sigma} F\, \mathrm{e}^{-r\tau}[N(d_2(S + F\, \mathrm{e}^{-r\tau})) - N(d_2(S))].$$

For $S \geqslant F \exp[-(r - \tfrac{1}{2}\bar{\sigma}^2)\tau]$, we have $d_2(S) \geqslant 0$, and we obtain

$$0 \geqslant e(S,t) \geqslant -[\bar{\sigma} F \mathrm{e}^{-r\tau}/\sqrt{(2\pi)}] \exp[-\tfrac{1}{2}(d_2(S))^2][d_2(S + F\, \mathrm{e}^{-r\tau}) - d_2(S)]$$

and, finally,

$$0 \geqslant e(S,t) \geqslant -[F\, \mathrm{e}^{-r\tau}/\sqrt{(2\pi\tau)}] \exp[-\tfrac{1}{2}(d_2(S))^2] \log[1 + (F\, \mathrm{e}^{-r\tau}/S)], \tag{32}$$

which corresponds to an extremely small error. It is smaller than any polynomial function of $1/S$.

This research was partly supported by the HEC Foundation and the Krueger Center for Finance at the Jerusalem School of Business.

Appendix A. *Proof of Proposition* 3.1

First, we prove that $\sigma(S,t)$ is solution of (11). Let us define $\psi(S,t) \equiv \log V(S,t)$. It then follows from the equity volatility function (7) that

$$\sigma(S,t) = \bar{\sigma}/S\psi_S. \tag{33}$$

From (10) and the definition of $\psi(S,t)$ it follows that

$$-\psi_t - \tfrac{1}{2}S^2\sigma^2(\psi_{SS}+\psi_S^2) - rS\psi_S + r = 0. \tag{34}$$

But from (33)

$$\psi_S = \bar{\sigma}/S\sigma, \quad \psi_{SS} = -\bar{\sigma}(\sigma + S\sigma_S)/S^2\sigma^2.$$

By substituting these expressions in equation (34), and rearranging terms we get

$$\psi_t = -\tfrac{1}{2}[(\bar{\sigma})^2 - \bar{\sigma}\sigma - \bar{\sigma}S\sigma_S] + r - (r\bar{\sigma}/\sigma).$$

Differentiating this last expression with respect to S yields

$$\psi_{tS} = \tfrac{1}{2}\bar{\sigma}\sigma_S + \tfrac{1}{2}\bar{\sigma}(\sigma_S + S\sigma_{SS}) + (r\bar{\sigma}\sigma_S/\sigma^2).$$

But also from (33) the partial differential of ψ_S is

$$\psi_{St} = -\bar{\sigma}\sigma_t/S\sigma^2.$$

Equating these two expressions for ψ_{tS} and ψ_{St} and rearranging terms, we finally obtain (11).

We now turn to the proof that $\tilde{\sigma}(V,t)$ is solution of (12) by following the same line of reasoning as for $\sigma(S,t)$. Define $\phi(V,t) \equiv \log S(V,t)$ and it then follows from (8) that

$$\tilde{\sigma}(V,t) = \bar{\sigma}V\phi_V. \tag{35}$$

From (3) and the definition of $\phi(V,t)$ it can be seen that

$$-\phi_t - \tfrac{1}{2}\bar{\sigma}^2V^2(\phi_{VV}+\phi_V^2) - rV\phi_V + r = 0. \tag{36}$$

But from (35) it follows that

$$\phi_V = \tilde{\sigma}/\bar{\sigma}V, \quad \phi_{VV} = (1/\bar{\sigma})\,[(\tilde{\sigma}_V/V) - (\tilde{\sigma}/V^2)].$$

If we substitute these expressions in (36), after simplifications, we obtain

$$\phi_t = -\tfrac{1}{2}\bar{\sigma}V\tilde{\sigma}_V + \tfrac{1}{2}\bar{\sigma}\tilde{\sigma} - \tfrac{1}{2}\tilde{\sigma}^2 - (r\tilde{\sigma}/\bar{\sigma}) + r.$$

Differentiating this expression with respect to V yields

$$\phi_{tV} = -\tfrac{1}{2}\bar{\sigma}V\tilde{\sigma}_{VV} - (r + \bar{\sigma}\tilde{\sigma})\,\tilde{\sigma}_V/\bar{\sigma}.$$

But, on the other hand, from (35)

$$\phi_{tV} = \tilde{\sigma}_t/\bar{\sigma}V. \qquad \square$$

Equating these two expressions for ϕ_{tV} yields (12).

Appendix B

In what follows we derive intermediate results supporting Theorem 3.1, which helps in assessing the order of magnitude of the approximation error. In Proposition B.1 we derive an explicit formula for $\tilde{\theta}(V,t)$ as a function of the terminal condition $S(V,T)$. Then in Proposition B.2 we derive some important convergence properties. From these results we can prove that the approximation error tends to zero when S becomes large.

Lemma B.1. *The solution of the Black–Scholes* PDE (3) *is*

$$S(V,t) = \frac{\exp(-r\tau)}{\sqrt{(2\pi\tau)}\,\bar{\sigma}} \int_{-\infty}^{+\infty} \exp\left(-\frac{1}{2}\frac{x^2}{\tau\bar{\sigma}^2}\right) S(V\exp[(r-\tfrac{1}{2}\bar{\sigma}^2)\tau]\exp x, T)\,\mathrm{d}x, \tag{37}$$

where $\tau = T - t$.

The proof follows by directly checking that the function $S(V,t)$, as defined explicitly by (37), is the solution of (3), and takes the value $S(V,T)$ at time T. Note also that by differentiating (37) with respect to V we obtain

$$S_V(V,t) = \frac{1}{\sqrt{(2\pi\tau)}\,\bar{\sigma}} \int_{-\infty}^{+\infty} \exp\left[-\frac{1}{2\tau\bar{\sigma}^2}(x-\tau\bar{\sigma}^2)^2\right] S_V(V\exp[(r-\tfrac{1}{2}\bar{\sigma}^2)\tau]\exp x, T)\,\mathrm{d}x. \tag{38}$$

Then we can derive the following result.

Proposition B.1.

$$\tilde{\theta}(V,t) = \frac{\exp(-r\tau)}{\sqrt{(2\pi\tau)}} \int_{-\infty}^{+\infty} \left(\frac{x}{\tau\bar{\sigma}^2}-1\right)\exp\left(-\frac{1}{2}\frac{x^2}{\tau\bar{\sigma}^2}\right) S(V\exp[(r-\tfrac{1}{2}\bar{\sigma}^2)\tau]\exp x, T)\,\mathrm{d}x. \tag{39}$$

Proof. From (38) we obtain

$$VS_V(V,t) = V\frac{1}{\sqrt{(2\pi\tau)}\,\bar{\sigma}}\exp(-\tfrac{1}{2}\bar{\sigma}^2\tau)\int_{-\infty}^{+\infty}\exp x\exp\left(-\frac{1}{2}\frac{x^2}{\tau\bar{\sigma}^2}\right) \times S_V(V\exp[(r-\tfrac{1}{2}\bar{\sigma}^2)\tau]\exp x, T)\,\mathrm{d}x,$$

which can be rewritten as

$$VS_V = \frac{\exp(-r\tau)}{\sqrt{(2\pi\tau)}\,\bar{\sigma}}\int_{-\infty}^{+\infty}\exp\left(-\frac{1}{2}\frac{x^2}{\tau\bar{\sigma}^2}\right)\frac{\mathrm{d}}{\mathrm{d}x}S(V\exp[(r-\tfrac{1}{2}\bar{\sigma}^2)\tau]\exp \mathrm{x}, T)\,\mathrm{d}x.$$

Integrating by parts, we get

$$VS_V = \frac{\exp(-r\tau)}{\sqrt{(2\pi\tau)}\,V\bar{\sigma}}\int_{-\infty}^{+\infty}\frac{x}{\tau\bar{\sigma}^2}\exp\left(-\frac{1}{2}\frac{x^2}{\tau\bar{\sigma}^2}\right)S(V\exp[(r-\tfrac{1}{2}\bar{\sigma}^2)\tau]\exp x, T)\,\mathrm{d}x,$$

which yields (39), recalling the definitions (13) for $\tilde{\theta}(V,t)$ and (8) for σ. □

By differentiating (39) we obtain

$$\tilde{\theta}_V(V,t) = \frac{\exp(-r\tau)}{\sqrt{(2\pi\tau)}\,V\tau\bar{\sigma}^2}\int_{-\infty}^{+\infty}\left(\frac{x^2}{\tau\bar{\sigma}^2}-1-x\right)\exp\left(-\frac{1}{2}\frac{x^2}{\tau\bar{\sigma}^2}\right) \times S(V\exp[(r-\tfrac{1}{2}\bar{\sigma}^2)\tau]\exp x, T)\,\mathrm{d}x. \tag{40}$$

An alternative formula for the derivative $\tilde{\theta}_V(V,t)$ is

$$\tilde{\theta}_V(V,t) = \frac{1}{\sqrt{(2\pi\tau)}}\frac{1}{\tau\bar{\sigma}^2}\int_{-\infty}^{+\infty}x\exp\left(-\frac{1}{2}\frac{x^2}{\tau\bar{\sigma}^2}\right)S_V(V\exp[\tfrac{1}{2}\bar{\sigma}^2+r]\tau)\exp x, T)\,\mathrm{d}x. \tag{41}$$

Formula (41) follows by differentiating the alternative specification for $\tilde{\theta}(V,t)$, namely

$$\tilde{\theta}(V,t) = \frac{\exp[-(r+\tfrac{1}{2}\bar{\sigma}^2)\tau]}{\sqrt{(2\pi\tau)}\,\tau\bar{\sigma}^2}\int_{-\infty}^{+\infty}x\exp(-x)\exp\left(-\frac{1}{2}\frac{S^2}{\tau\bar{\sigma}^2}\right) \times S(V\exp[(\tfrac{1}{2}\bar{\sigma}^2+r)\tau]\exp x, T)\,\mathrm{d}x, \tag{42}$$

which stems from (39) after we operate the change of variable $x \to x+\tau\bar{\sigma}^2$ in the integral. We can now derive convergence properties that are useful in the derivation of Theorem 3.1, where we show that the approximation error tends to zero when S becomes large.

Proposition B. 2. *Suppose there exists a constant a, such that for any constant M,*

$$\begin{aligned} &V(S_V(V,T)-a)\to 0 \quad \text{as} \quad V\to\infty, \\ &|V(S_V(V,T)-a)|\leqslant M. \end{aligned} \tag{43}$$

Then, for a convenient function of time $K(t)$

$$\begin{aligned} &V\tilde{\theta}_V(V,t)\to 0 \quad \text{as} \quad V\to\infty, \\ &|V\tilde{\theta}_V(V,t)|\leqslant K(t), \forall\, V. \end{aligned} \tag{44}$$

Proof. Using (41) and the fact that we can add any constant a in the integral without changing its value

$$\left(\text{indeed} \int_{-\infty}^{+\infty} x \exp\left(-\frac{1}{2}\frac{x^2}{\tau\bar{\sigma}^2}\right) \mathrm{d}x = 0\right),$$

we can write

$$V\tilde{\theta}_V(V,t) = \frac{1}{\sqrt{(2\pi\tau)}}\frac{1}{\tau\bar{\sigma}^2}\int_{-\infty}^{+\infty} x\exp\left(-\frac{1}{2}\frac{x^2}{\tau\bar{\sigma}^2}\right) V\{S_V(V\exp[(\tfrac{1}{2}\bar{\sigma}^2+r)\,\tau]\exp x, T)-a\}\,\mathrm{d}x.$$

Take a sequence $V^n\to\infty$ as $n\to\infty$, then for any fixed x, it follows from assumptions (43) that

$$V^n\{S_V(V^n\exp[(\tfrac{1}{2}\bar{\sigma}^2+r)\,\tau]\exp x, T)-a\}\to 0.$$

The integrand is in modulus majorized by the fixed integral function

$$\tilde{K}(t)\,|x|\exp(-x)\exp[-\tfrac{1}{2}(x^2/\tau\bar{\sigma}^2)],$$

where $\tilde{K}(t) = M\exp[-(\tfrac{1}{2}\bar{\sigma}^2+r)\,\tau]$. Hence, (44) follows simply from Lebesgue's theorem. The boundedness property is an immediate consequence of the second of assumptions (43). □

References

Bensoussan, A., Crouhy, M. & Galai, D. 1993 Black–Scholes approximation of complex option values: the case of European compound call options and equity warrants. Groupe HEC (Working paper).

Black, F. & Scholes, M. 1973 The pricing of options and corporate liabilities. *J. Political Econ.* **81**, 637–54.

Crouhy, M. & Galai, D. 1994 The interaction between the financial and investment decisions of the firm: the case of issuing warrants in a levered firm. *J. Banking Finance.* (In the press.)

Galai, D. & Masulis, R. 1976 The option pricing model and the risk factor of stock. *J. Financial Econ.* **3**, 53–81.

Merton, R. C. 1973 Theory of rational option pricing. *Bell J. Econ. Management Sci.* **4**, 141–83.

Discussion

A. D. Wilkie (*Watsons, Reigate, U.K.*). A system in which the share price is known was described, presumably the price for the debt is not known, and the value of the company is not known. It does not seem possible to rely on arbitrage arguments between the price of the share and the value of the company. You concentrate on the relations between the standard deviations. But does your argument also not apply to the mean rate of growth of the value of the company? If arbitrage is possible, then the mean rate of growth of the underlying security does not enter the formula for the valuation of an option, and the argument of using an equivalent martingale measure leads one to using the risk free rate. But in this case don't you have to value the share according to the underlying rate of growth of the company, and not at the risk free rate?

M. Crouhy. In our model there is only one source of exogenous uncertainty which is the value of the firm's assets. It drives the dynamics of all the financial claims issued by the firm whether they are equity, debt, convertible bonds or warrants. This source of risk is hedgeable, and therefore it is perfectly legitimate to use the risk neutral valuation framework. In fact by assuming in equation (1), which specifies the stochastic process followed by the assets of the firm, that the drift term is the risk free interest rate, we are already working with the risk-neutral probability measure, Q. Under Q, dz is a brownian motion and discounted security prices are Q-martingales.

M. A. H. Dempster (*University of Essex, U.K.*). It would be interesting to explore the implications of replacing the assumption of a deterministic functional relation $S(t) = S(V(t), t)$, between a realized asset value V of the firm and the corresponding stock price S, by the stochastic differential equation

$$\mathrm{d}S = f(V(t), t) + \mathrm{d}W,$$

where W is an independent Wiener process representing ideosyncratic random effects on the firm's asset value not known to – and hence not valued by – the market. This would make the valuation problems treated partly observed control problems.

9

The general mean-variance portfolio selection problem

HARRY M. MARKOWITZ

This paper states the 'general mean-variance portfolio analysis problem' and its solution, and briefly discusses its use in practice.

---

1. The problem

We consider n securities whose returns $r' = (r_1, \ldots, r_n)$ during the forthcoming period have expected values $\mu' = (\mu_1, \ldots, \mu_n)$ and a covariance matrix $C = (\sigma_{ij})$. An investor is to select a portfolio $X' = (X_1, \ldots, X_n)$. The return $R = r'X$ on the portfolio has expected value and variance, respectively,

$$E = \mu' X, \quad V = X'CX. \tag{1a, b}$$

The portfolio is to chosen subject to constraints

$$AX = b, \quad X \geqslant 0, \tag{1c, d}$$

where A is $m \times n$ and b is $m \times 1$. Thus, non-negative X_i are to be chosen subject to $m \geqslant 1$ linear inequalities.

A portfolio is *feasible* if it satisfies (1c) and (1d). An EV combination is feasible if it is the E and V of a feasible portfolio. A feasible EV combination (E_0, V_0) is *inefficient* if there is another feasible EV combination (E_1, V_1) such that either

$$\text{(i)}\ E_1 > E_0 \quad \text{and} \quad V_1 \leqslant V_0$$

or

$$\text{(ii)}\ V_1 < V_0 \quad \text{and} \quad E_1 \geqslant E_0.$$

A feasible EV combination is efficient if it is not inefficient. A feasible portfolio is efficient or inefficient in accordance with its EV combination.

It is not sufficient to require only condition (i) or condition (ii) in the definition of inefficiency, nor to define an efficient EV combination as one which maximizes E for given V and minimizes V for given E. Examples can be constructed of feasible EV combinations which meet the latter two requirements but are nevertheless inefficient as previously defined.

Since C is a covariance matrix it is positive semi-definite. We do not require it to be positive definite because C is singular in some important applications. We will see an example below. If C is singular, there may be more than one efficient portfolio which has a given efficient EV combination. We define a 'complete, non-redundant' set of efficient portfolios as one which contains one and only one efficient portfolio for each efficient EV combination.

The portfolio analysis problem is to determine

1. whether constraints (1c) and (1d) are feasible and, if they are, calculate
2. the set of all efficient EV combinations and
3. a complete, non-redundant set of efficient portfolios.

Mathematical Models in Finance. Edited by S.D. Howison, F.P. Kelly and P. Wilmott.
Published in 1995 by Chapman & Hall, London. ISBN 0 412 63070 2

Examples exist of portfolio selection problems with feasible portfolios but no efficient portfolios. This is possible (but not necessary) when E is unbounded and C singular. Excluding this case, every feasible portfolio selection problem has a piecewise parabolic set of efficient EV combinations, with a finite number of pieces. For every such problem there exists a piecewise linear complete, non-redundant set of efficient portfolios. For these assertions to apply with the generality stated, we define 'piecewise' to include a single 'piece' (line or parabolic segment) or only a point. One piece (line segment and parabolic segment) may be unbounded in one direction.

We will consider the computation of these efficient sets after we further consider problem definition and application.

2. Application

It might seem that we could gain some generality by allowing linear inequalities ($\geqslant$, $\leqslant$) in (1c) or allowing some or all variables to be negative. This is not the case. Specifically, given any mean-variance portfolio analysis problem whose constraints permit some or all variables to be negative and permits some or all linear constraints in (1c) to be (weak) inequalities, there is a problem in the standard form of (1c) and (1d) that has the same answer. That is, it has the same set of efficient EV combinations (even the same set of feasible EV combinations) and, given the complete, non-redundant set of efficient portfolios for the equivalent problem it is easy to determine this for the original problem. (See chapter 2 of Markowitz (1987) for details.)

Thus, as far as applications are concerned, we think of the general mean-variance portfolio selection problem as one of finding mean-variance efficient portfolios in variables that may or may not be required to be non-negative, which are subject to one or more (actually, zero or more) linear equalities or weak inequalities. (Also, certain nonlinear constraints can be approximated, as in linear programming.)

Examples of such linear constraints are the budget constraint, a 'turnover constraint' which limits the amount by which the new portfolio may differ from the previous one, and upper or lower bounds on the amount invested in a security or an industry. A special case is an 'exogenous asset' whose amount is fixed in the portfolio. The exogenous asset may, for example, be a random source of income other than the return on securities. For a given E, the variance minimizing portfolio as a whole, including the exogenous asset, depends on the covariances between the exogenous asset and the securities whose amounts are to be selected.

The exogenous asset may be a state variable for which the investor wishes to seek or avoid correlation of portfolio return. The single period mean-variance analysis should be thought of as an approximation to the single period derived utility maximization which is optimal within a many-period investment game. If the derived utility function depends on state variables other than end-of-period wealth, the mean variance approximation may use exogenous assets. (See chapter 3 of Markowitz (1987), starting from the section 'Why mean and variance'.)

We noted that one can convert a model with inequality constraints to an equivalent one with equality constraints. This involves introducing slack variables, as in linear programming. Since slack variables have zero variance, their presence makes C singular. Singular C is no problem for the 'critical line' algorithm described in the next section.

3. Computation

Every feasible general portfolio selection problem has a solution of the following nature. Each piece of the piecewise linear set of efficient portfolios has a set $\text{IN} \subset \{1, \ldots, n\}$ of 'in' securities. The others are 'out'. Obtain A_{IN} and μ_{IN} by setting $a_{ki} = 0$ and $\mu_i = 0$ if $i \in \text{OUT}$. Obtain C_{IN} by setting $\sigma_{ij} = \delta_{ij}$ ($= 1$ if $i = j$, otherwise $= 0$) if either i or j is out.

If

$$M_{\text{IN}} = \begin{pmatrix} C_{\text{IN}} & A'_{\text{IN}} \\ A_{\text{IN}} & 0 \end{pmatrix} \tag{2a}$$

is non-singular, then the solution to

$$M_{\text{IN}} = \begin{pmatrix} X \\ \lambda \end{pmatrix} = \begin{pmatrix} 0 \\ b \end{pmatrix} + \begin{pmatrix} \mu_{\text{IN}} \\ 0 \end{pmatrix} \lambda_E, \tag{2b}$$

i.e.
$$\begin{pmatrix} X \\ \lambda \end{pmatrix} = M_{\text{IN}}^{-1} \begin{pmatrix} 0 \\ b \end{pmatrix} + M_{\text{IN}}^{-1} \begin{pmatrix} \mu_{\text{IN}} \\ 0 \end{pmatrix} \lambda_E = \alpha_{\text{IN}} + \beta_{\text{IN}} \lambda_E, \tag{2c}$$

is referred to as a 'critical line'.

Typically, most critical lines contain no efficient portfolios. If an efficient point exists on a critical line, then there is an interval of efficient portfolios, namely, the solution to (2) for

$$\lambda_E \in [\lambda_{\text{LOW}}, \infty) \quad \text{or} \quad \lambda_E \in [\lambda_{\text{LOW}}, \lambda_{\text{HI}}]. \tag{3a, b}$$

This is referred to as the 'efficient segment' of the critical line.

In particular, there exists a complete, non-redundant set of efficient portfolios which consists of the efficient segments of $K \geqslant 1$ critical lines. These critical lines are efficient for

$$\lambda_E \in [\lambda_{\text{LOW}}^1, \infty), \quad \lambda_E \in [\lambda_{\text{LOW}}^2, \lambda_{\text{HI}}^2], \ldots, \lambda_E \in [0, \lambda_{\text{HI}}^K],$$

where $\lambda_{\text{HI}}^{k+1} = \lambda_{\text{LOW}}^k$ and $\lambda_{\text{HI}}^k \geqslant \lambda_{\text{LOW}}^k$.

IN^{k+1} differs from IN^k by the addition or deletion of one member. In the relatively easy to explain 'non-degenerate' case, to be defined below, $\lambda_{\text{HI}}^k > \lambda_{\text{LOW}}^k$ for $k = 1, \ldots, K$.

Define $\eta' = (\eta_1, \ldots, \eta_n)$ by

$$\eta = (C_{\text{IN}}, A'_{\text{IN}}, -\mu_{\text{IN}}) \begin{pmatrix} X \\ \lambda \\ \lambda_E \end{pmatrix}. \tag{4}$$

η_i is the partial derivative of a lagrangian with respect to X_i. Substituting (2c) into (4) we obtain η as a linear function of λ_E

$$\eta = \gamma_{\text{IN}} + \delta_{\text{IN}} \lambda_E. \tag{5}$$

A sufficient condition for a point on a critical line to be an efficient portfolio is that

$$X \geqslant 0, \quad \eta \geqslant 0, \quad \lambda_E > 0. \tag{6a–c}$$

A portfolio satisfying (6a), (6b) and $\lambda_E = 0$ may or may not be efficient. We return to this point below.

The critical line algorithm computes a complete, non-redundant set of efficient portfolios. The computation is simplest when the problem is non-degenerate (still to be defined) and feasible E is bounded above. Part of the non-degeneracy assumptions (to be relaxed below) are that (a) there is a unique feasible portfolio which maximizes E, and (b) the problem of maximizing E subject to ($1c$) and ($1d$) is non-degenerate in the sense defined in linear programming. In this case, the unique optimum solution has exactly m variables – the 'basis variables' – with $X_i > 0$. These basis variables are the first IN set. With IN_1 thus defined, M_{IN} is non-singular, and M_{IN}^{-1} is easy to compute. The critical line

$$\begin{pmatrix} X \\ \lambda \end{pmatrix} = \alpha + \beta\lambda_E, \quad \eta = \gamma + \delta\lambda_E, \tag{7a, b}$$

satisfies ($6a, b$) and $\lambda_E \geqslant 0$ for all

$$\lambda_E \in [\lambda^1_{\text{LOW}}, \infty),$$

where λ_{LOW} is the largest λ_E below which one of the three conditions becomes false; that is, it is the first (i.e. largest) λ_E at which $\eta_i \downarrow 0$ for some OUT i or $\lambda_E \downarrow 0$. With subsequent IN sets we also have the possibility that $X_i \downarrow 0$ for some IN i; but with IN_1, i.e. on the first critical line, X is constant and equal to the E maximizing portfolio. Whereas ($6a, b$) and $\lambda_E = 0$ do not in general assure an efficient portfolio, if $\lambda^1_{\text{LOW}} = 0$ then the portfolio with maximum E is also the efficient portfolio with minimum V, and is thus the entire complete non-degenerate set of efficient portfolios. We next consider the case of $\lambda^1_{\text{LOW}} > 0$.

The remaining non-degeneracy assumption is that, at each iteration k, a unique i determines $\lambda^k_{\text{LOW}} > 0$. In particular, in the first iteration a unique $\eta_{i^*} \downarrow 0$ first. IN_2 is the same as IN_1 except for the addition of i^*. The new M_{IN} associated with IN_2 is guaranteed to be non-singular.

The fine print in this guarantee says that the statement would be true if we performed calculations with unlimited precision. This could actually be done by storing the numerator and denominator of all results as unlimited, but always finite, integers; since only rational operations are performed. In practice, floating point arithmetic is used, round-off errors occur, and the aforementioned and other 'guarantees' are no longer certain. As a matter of fact, the computation succeeds, including satisfying conditions which assure optimality, most of the time even for $n \geqslant 1000$. When it fails, some rescaling or a practically equivalent restatement of the problem usually, perhaps always, succeeds. The fact that the algorithm usually works does not diminish the desirability of understanding the round-off error problem much better than we do. From this point on we ignore the round-off error problem and, in effect, assume that computations are performed with unlimited precision.

Given our current non-degeneracy assumptions, X_{i^*} will increase as λ_E is reduced below λ^1_{LOW}. Conditions ($6a, b$) and $\lambda_E \geqslant 0$ will hold for

$$\lambda_E \in [\lambda^2_{\text{LOW}}, \lambda^2_{\text{HI}}],$$

where $\lambda^2_{\text{HI}} = \lambda^1_{\text{LOW}}$, $\lambda^2_{\text{LOW}} < \lambda^2_{\text{HI}}$ and $\lambda^2_{\text{LOW}} \geqslant 0$. λ^2_{LOW} is the largest value of $\lambda_E < \lambda^2_{\text{HI}}$ below which one of the three conditions will be violated. If $\lambda^2_{\text{LOW}} = 0$ then the point on the second critical line at $\lambda_E = 0$ is the efficient portfolio with minimum variance. (When C is singular, there may be other feasible portfolios with the same V but lower E.)

In case $\lambda^2_{\mathrm{LOW}} > 0$, one of our above-stated non-degeneracy assumptions assures us that one and only one i IN will satisfy

$$\alpha_i + \beta_i \lambda^2_{\mathrm{LOW}} = 0, \tag{8a}$$

or i OUT will satisfy

$$\gamma_i + \delta_i \lambda^2_{\mathrm{LOW}} = 0. \tag{8b}$$

($X_i = 0$ for i OUT and $\eta_i = 0$ for i IN at all points on the critical line.)

IN_3 differs from IN_2 by the deletion of the i^* that satisfies (8*a*) or the addition of the i^* that satisfies (8*b*). In either case, the new M_{IN} will be non-singular (even if C is singular). In case (8*a*), η_i will increase as λ_E is reduced below λ^2_{LOW}; in case (8*b*), X_i will increase as λ_E decreases.

Our discussion of IN_2, its corresponding critical line and their relation to IN_3 and its critical line, illustrates the general case. The same relations hold for IN_k and between IN_k and IN_{k+1}. The computation stops when $\lambda_E = 0$ is reached. This must happen in a finite number of steps, since the same IN set cannot appear twice.

4. Degeneracy and other problems

Now we tie up loose ends. First, we note that certain difficulties are handled for us when we use George Dantzig's simplex algorithm for solving the linear programming problem of maximizing E subject to (1*c*) and (1*d*). (See Dantzig (1963) for details concerning the simplex algorithm, cycling in linear programming, etc.) If the model is unfeasible, this is determined by 'phase I' of the simplex calculation, and we are so advised. If the rank of A is less than m, phase I provides an equivalent model which is not rank deficient.

Phase II informs us if feasible E is unbounded or if the E maximizing solution is not unique or is degenerate in the sense that some variable in the basis has $X_i = 0$. It also informs us as to which non-basic activities (columns of A) have 'zero profitability', i.e. have the partial derivative of the appropriate lagrangian equal to zero.

It has been shown that degenerate linear programming problems can cycle; that is, a sequence of iterations can occur in which variables enter and leave the basis but do not change value, and in which a given basis repeats itself. If the same (non-random) rule is followed to decide which non-basic variable with positive profit is to go into the basis, and which basic variable with zero value is to go out, then the once repeating basis will repeat infinitely often.

The problem of modifying the simplex algorithm so it is guaranteed not to cycle can be solved along the following lines. In principle, for sufficiently small positive ϵ, adding certain powers of ϵ to each b_i will produce a still feasible linear program which is not degenerate and has almost the same answer as the original linear program. The answer to this perturbed problem approaches the original answer as $\epsilon \downarrow 0$. It is not necessary to actually add these powers of ϵ, since one can calculate the sequence of bases which would occur for any sufficiently small positive ϵ.

In practice, the simplex algorithm never (or perhaps hardly ever) cycles, except in problems especially constructed to show that cycling is possible. In other words, if one ignores the existence of cycling and arbitrarily breaks ties for which variable goes in or out of the basis, then usually no problem is encountered.

It is not known whether the critical line algorithm will cycle if two or more η_i and/or X_i go to zero simultaneously and the critical line algorithm chooses one to enter or leave the IN set by an arbitrary rule. To my knowledge, in practice such cycling has never happened. However, in case it ever does, one can build a version of the critical line algorithm that is guaranteed not to cycle, by, in effect, adding suitable powers of ϵ to the b_i and μ_i. The sequence of IN sets is the same for all sufficiently small positive ϵ. As in linear programming, it is not necessary to actually add these powers of ϵ to figure out the sequence of IN sets. A solution to the original problem can be determined from the IN sets of the perturbed problem.

A problem with unbounded E can be reduced to one with bounded E by adding the constraint

$$\mu' X \leqslant E_0. \tag{9}$$

The sequence of IN sets is the same for all sufficiently large E_0. It is not necessary to actually add constraint (9) to determine this sequence of IN sets. From them, the solution to the original problem can be inferred. The solution includes one unbounded piece (line segment in portfolio space, parabolic segment in EV space) which is efficient for

$$\lambda_E \in [\lambda^1_{\text{LOW}}, \infty).$$

This completes our outline of the solution to the general portfolio problem for all possible inputs.

References

Dantzig, G. B. 1963 *Linear programming and extensions*. Princeton University Press.

Markowitz, H. M. 1956 *The optimization of a quadratic function subject to linear constraints. Nav. Res. Logistics Q.* vol. III.

Markowitz, H. M. 1987 *Mean-variance analysis in portfolio choice and capital markets*. Oxford: Basil Blackwell.

Discussion

R. Lacey (*Derivative Investment Advisers Ltd, U.K.*). How far can transaction costs analysis, described by Professor M. H. A. Davis and Dr P. Wilmott (this Volume), be incorporated into the portfolio selection method optimization module?

H. M. Markowitz. Transaction costs can be incorporated exactly if they are proportional to change in position; see Markowitz (1987). For an approximate solution when costs are linear but not proportional, see Perold (1984).

J. Plymen (*Ruislip, U.K.*). Consider the investment scene in the early 1960s when the Markowitz principles were developed. Investment statistics were rudimentary, with long term share indices confined to prices without any dividend record. Computers were too slow and expensive for any elaborate analysis. Mean-variance analysis with its crude one factor input was the only scientific technique available. Although investment inputs were developed using fundamental multifactor betas, this was only a small improvement.

United Kingdom actuaries adopted a different approach. In 1962 they developed the *Financial Times* Actuaries 500 share index. Next they set up equity market models that compare individual share performance with that of the index, obtaining a relative price ranking. (Models by Weaver & Hall, Hempsted and Clarkson have been published in actuarial journals.) With this pricing ability, portfolios are monitored at regular intervals, selling dear shares for cheap ones. This continual programme reduces risk and improves performance.

Mean-variance analysis based on more sophisticated models for performance and using semi-variance rather than mean-variance may have practical value. Mathematicians interested in finance could concentrate on: (i) actuarial market models; (ii) analytical techniques for various forms of derivatives; (iii) mathematical use of semi-variance rather than variance.

M. A. H. Dempster (*University of Essex, U.K.*). The type of index tracking portfolio management policy advocated by Mr Plymen can lead to inefficiencies with respect to any attitude-to-risk criterion, including mean-variance. As pointed out by Hodges (this Volume) the appropriate criterion for portfolio management depends on whose preferences – fund managers or ultimate beneficiaries – it embodies. In any event Professor Markowitz's recent practical portfolio experience with sophisticated mean-semi-variance methods is impressive.

R. G. Tompkins (*Kleinwort Benson Investment Management, U.K.*). In applying a mean-variance framework to emerging markets we find ridiculous results. That is, these markets have extraordinarily high historical returns and extremely low risk. This is counter-intuitive as we know how risky these markets are. We have found data series distributions to be extremely skewed and lepto Kurtic. It would be useful to expand the mean-variance framework in portfolio management to include the third and fourth moments. Perhaps this approach could be applied to the inclusion of contingent claims (such as options) into estimating optional portfolios.

Additional references

Clarkson, R. S. & Plymen, J. 1988 Improving the performance of equity portfolios. *J. Inst. Actuaries* **115**.

Perold, A. 1984 Large-scale portfolio optimization. *Man. Sci.* **30**, 1143–1160.

General discussion

R. S. Clarkson (*Dalserf, U.K.*). What is the best way forward for the theory of finance and the role that mathematics can play?

To apply the mathematical methods discussed we need the best possible understanding of the 'hidden variables' behind security prices. Professor Ziemba's comments on how errors in expected values in portfolio selection models are far more serious than errors in covariances confirm this. But how should we make these estimates?

We have two main choices. We can – as suggested by Edgar Peters in his book *Chaos and order in the capital markets* – invoke chaos theory and try to find ever more complex mathematical relationships between security prices and other factors. This approach is unlikely to be of much assistance. The alternative is to go back to the basics of understanding the human mind and find a simple model of human behaviour to replace the expected utility maxim that is a central feature of economists' definitions of rational behaviour.

An illuminating description of human behaviour under conditions of uncertainty and risk is chapter 12 of Keynes' *The general theory of employment, interest and money*, titled 'The state of long-term expectation'. Keynes says that 'confidence' is of far greater importance than generally realized. The downside measure of financial risk I devised encapsulates this 'confidence' in mathematical terms. Three comments may be of interest.

First the unthinking use of variance of return as the measure of risk has seriously weakened the credibility of the current theory of finance.

Second, it was only after detailed examination of some resultant contradictions – in particular the 'reasonable behaviour' cited by the French economist Professor Allais – that Professor Markowitz adopted the expected utility maxim for portfolio selection purposes. His justification is:

> 'This writer believes that the arguments in favor of the expected utility maxim are quite convincing, especially for its application in areas such as portfolio selection. The maxim has to be stretched, perhaps intolerably, to apply to the making of decisions in which surprise and the fun of gambling are important motivations. These, however, are not important objectives for the direction of a machine in the allocation of large amounts of other people's money.'

I wholeheartedly agree in the case of portfolio selection. However, for individual investors a generalized version of semi-variance is required to reflect certain other background factors such as the individual's current wealth and degree of aversion to risk.

My third point relates to whether Markowitz's portfolio selection approach is an outdated methodology. Having derived my risk measure it was interesting to see what option prices resulted. I worked out put and call option prices for periods from one month to one year and calculated the corresponding prices using the Black–Scholes model. The two sets of values were, for all practical purposes, identical. If a generalization of the measure of risk now used by Professor Markowitz

Mathematical Models in Finance. Edited by S.D. Howison, F.P. Kelly and P. Wilmott.
Published in 1995 by Chapman & Hall, London. ISBN 0 412 63070 2

can generate the Black–Scholes values without the highly unrealistic assumptions needed by Black and Scholes themselves, it seems obvious that the Markowitz, not the Black–Scholes, approach, is more fundamental in terms of describing our complex financial world.

S. D. HODGES (*Warwick University, U.K.*). A case-study which illustrates the best kind of interaction between finance and mathematics relates to work on transactions costs (see Merton 1971; Constantinides 1986; Davis & Norman 1993). A. Neuberger and I were aware of this literature and we independently formulated option hedging as an optimal control problem. We realized that if we used exponential utility we would reduce the number of state variables from three (plus time) to two, making the problem computationally feasible. Hodges & Neuberger (1989) gave the correct formulation, the qualitative aspects of the solution and the results of our numerical computations.

Mathematicians have important contributions to make in adding rigour to complicated analysis (particularly where there is a chance that our intuition has misled us), in helping to solve particular problems (analytically or numerically) and probably in drawing out generalizations.

However, there are dangers. Mathematicians untrained in finance do not possess skills in formulating problems in appropriate and useful ways. It may not be helpful to produce exactly the weakest sufficient conditions for something to work, and in notation that few finance people can follow, when the existing assumptions are adequate to justify a result.

In conclusion, it is important that mathematicians interested in finance should become acquainted with the academic finance literature and also work with academic finance specialists.

A. J. G. CAIRNS (*Heriot-Watt University, U.K.*). Derivative pricing is largely driven by short term views of individual stocks or the market as a whole, and by how to increase profit. Many insurance problems, on the other hand, require the actuary to take long term effects into account and to look at strategic asset allocation rather than individual stock selection, which is rightly left to the specialist investment managers. There is a similarity here with modern portfolio theory, both being concerned with efficient and optimal allocation of assets. Insurance problems are, however, complicated by the need to take into account the underlying, varying liabilities and a long term rather than a short term view.

There is considerable room for development of such asset/liability modelling techniques. For example, derivatives are rarely included in such models. Such assets are being used increasingly for the purpose of risk reduction and it seems appropriate to incorporate them in our asset/liability models. Models such as Black–Scholes are a natural starting point, but we need to reconcile the theoretical incompatibility of, for example, geometric brownian motion, with the long-term, autoregressive interest rate models commonly used in actuarial work (for example, Wilkie 1987). These differences may not be significant, however, since autoregressive models can often be well approximated by brownian motion in the short term.

It is necessary to consider the robustness of results relative to parameter variation or uncertainty and, equally importantly to model choice. Related to this is the need to know which aspects of a model largely determine the outcome. The concept of such primary components has been considered in detail by Cairns (1991, 1993) and by

Brown and Schaefer (in this Volume), who show that the term structure of interest rates can be safely reduced to a two factor model.

T. Leach (*Assured Asset Management plc, Isle of Man, U.K.*) It would be useful to see research from the standpoint of the international investor. Current research for example on the Dow Jones or S&P 500 is done from the simplistic perception of a dollar-based investor for whom these indices are a true reflection of American equity performance. Adjustment for a currency index is necessary to show true performance from the perspective of the international investor and this adjustment assists in interpreting market movements which are affected substantially by international currency cash flows. It would also be helpful for analysis to use more detailed data such as hourly, ten minute, or tick by tick data as, with the reduction in transaction costs through the use of derivatives, predictive systems used in taking advantage of short term market movements will be more in demand.

J. W. E. Mariathasan (*Commercial Union Investment Management Ltd., U.K.*). Given that the asset management activities of a general insurance company encompass not only insurance funds, but management of a range of external funds such as pension funds and unit trusts, the potential uses of derivatives could, in theory, be much wider than for most more specialized fund managers. However, most derivative brokers do not understand the objectives and constraints within general insurance companies and therefore waste energy attempting to sell inappropriate products.

I would make a clear distinction between the use of futures contracts and forwards, which are well established and very useful tools for switching between asset classes and managing cashflows, and the use of option-type instruments, which introduce a much more structured risk return profile into the general pot of assets and liabilities. There are three areas where options can be used in investment management for a general insurance company:

(i) Use by individual fund managers of stock options in place of stock purchases and use of index options in general management of portfolios.

(ii) Transformation of wholesale products into retail products through guaranteed return funds etc.

(iii) Use of options at balance sheet management level to manage the group balance sheet, for example through the purchase of put options for protection during periods perceived to be particularly volatile.

In general, the market for equity derivatives is vastly overbroked in the UK. Numerous conversations with the major derivative brokers in the city indicate that in the UK at least, there is a lot less business in this area than media attention indicates. Derivative houses have targeted too much sales effort at fund manager level. Fund managers in traditional houses tend to be steeped in the analysis of stocks for determination of relative values and, as a result, are not very keen on the added complication of dealing in derivatives of the stocks which have a complex risk/return distribution. They are more comfortable adjusting exposures to stocks through adjusting the amount of the stock held rather than through the use of a derivative instrument. Their problems lie, not in finding a mechanism for turning a view on a stock into a profitable trade, but rather, in what their view on a stock should actually be.

Transforming wholesale products into retail products has proved very popular during the last few years, as seen by the success of numerous guaranteed return funds. Highly structured option products do have a place here. However, the reduction in interest rates in the UK, together with those that have already appeared, has reduced the potential for further ones, although there may be more appearing over the next year or so.

Balance sheet management is what really excites derivative brokers because insurance companies who wish to protect any balance sheet exposures have to deal in very large nominal sizes. For such institutions, the risk/return profile may well not be defined in high detail. Given then, the requirement to deal in large sizes to adjust the risk/return profile in a broad, rather than a highly structured sense, the institution is driven towards the use of highly liquid futures markets to adjust risk exposures, rather than through dealing in small amounts of esoteric highly structured over-the-counter option products. The latter may be very interesting and profitable for the investment bank selling the product, but are not of greater use for balance sheet management of a large institution.

To summarize, large institutions find straightforward exchange traded futures contracts of far greater use than the more esoteric highly structured over-the-counter option products.

References

Cairns, A. J. G. 1991 Model fitting and projection of the AIDS epidemic. *Math. Biosciences* **107**, 451–489.

Cairns, A. J. G. 1993 Primary components of epidemic models. In *Epidemic models: their structure and relation to data*, (ed. D. Mollison). Cambridge University Press.

Constantinides, G. M. 1986 Capital market equilibrium with transaction costs. *J. Political Econ.* **94**, 842–862.

Davis, M. H. A. & Norman, A. R. 1990 Portfolio selection with transaction costs. *Math. Operations Res.* **15**, 676–713.

Hodges, S. D. & Neuberger, A. 1989 Optimal replication of contingent claims under transactions costs. *Rev. Futures Markets* **8**, 224–242.

Merton, R. C. 1971 Optimum consumption and portfolio rules in a continuous-time model. *J. Econ. Theory* **3**, 373–413.

Wilkie, A. D. 1987 Stochastic investment models – theory and applications. *Insurance: Math. Econ.* **6**, 65–83.

10

On a free boundary problem that arises in portfolio management

Stanley R. Pliska and Michael J.P. Selby

We study a model for the optimal management of a portfolio when there are transaction costs proportional to a fixed fraction of the portfolio value. The risky securities are modelled as correlated geometric brownian motions. There is a riskless bank account and the aim is to maximize the long-run growth rate. It is known that the optimal trading strategy is characterized by the solution of a certain partial differential equation free boundary problem. This paper explains how to transform this free boundary problem for the case of three securities into a much simpler one that is feasible to solve with numerical methods.

---

1. Introduction

In a closely related paper, Morton & Pliska (1993) introduce and develop a model for the optimal management of a portfolio with fixed transaction costs. Their model features a savings account with a constant interest rate r and m risky stocks whose prices $Z_t^k, k = 1, \ldots, m$, are correlated geometric brownian motions, that is,

$$\mathrm{d}Z_t^k = Z_t^k\left(\mu_k\,\mathrm{d}t + \sum_{j=1}^{m} \lambda_{kj}\,\mathrm{d}W_t^j\right), \quad k = 1, \ldots, m,$$

where μ is an m vector of appreciation rates, λ_{ij} is the (i,j) entry in the m by m matrix Λ, and $W^1, \ldots, W^m$ are independent brownian motion processes. Thus the time t price Z_t^k of stock k has a log-normal distribution with mean $Z_0^k \exp(\mu_k t)$. It is assumed that the variance–covariance matrix $M \equiv \Lambda\Lambda'$ is of full rank and that all the components of the vector $(\Lambda\Lambda')^{-1}(\mu - r1)$ are strictly positive and sum to less than one.

The portfolio manager starts with initial capital V_0 and strives to maximize the long-run, asymptotic growth rate,

$$\lim_{T\to\infty} \inf[E(\ln V_T)/T],$$

where V_T is the value of (that is, the amount of money in) the portfolio at time T. No money can be added to or withdrawn from the portfolio (except for transaction costs paid to the broker), there is no short-selling of stocks and there is no borrowing of funds at the interest rate r.

The portfolio manager is free to use a very general, non-anticipative trading strategy governing the allocation of funds between the individual stocks and the savings account, but each time a transaction occurs, that is, each time funds are shifted between two or more stocks or between the stocks and the savings account,

Mathematical Models in Finance. Edited by S.D. Howison, F.P. Kelly and P. Wilmott.
Published in 1995 by Chapman & Hall, London. ISBN 0 412 63070 2

a transaction cost is incurred equal to the fraction $(1-\alpha)$ times the current value of the whole portfolio (here $0 < \alpha < 1$, and normally α is close to 1). Aside from the portfolio's value, this transaction cost does not depend on the number of, the prices of, or the positions in the stocks that are involved in the transaction at the point in time when the transaction occurs. Hence if a transaction occurs at time T when the value of the portfolio is V_{T-}, then the amount $(1-\alpha)V_{T-}$ is paid to the stock broker and the portfolio continues with the new value $V_T = \alpha V_{T-}$.

Morton & Pliska (1993) show that the trading strategy that maximizes the asymptotic growth rate is fully described by an m vector b (whose components are strictly positive and sum to less than one) and a stopping time τ (which will be described below). The idea is very simple. The portfolio manager starts out with the initial funds allocated among the stocks according to the vector b, and the balance of the funds in the savings account. In other words, $b_k V_0$ equals the initial investment in stock $k, k = 1, \ldots, m$. The first transaction occurs according to the stopping rule τ, the amount $(1-\alpha)V_{\tau-}$ is paid to the stock broker and the remaining funds $V_\tau = \alpha V_{\tau-}$ are reallocated (that is, rebalanced) according to the vector b. Now $b_k V_\tau$ equals the investment in security k, and this cycle then repeats itself indefinitely.

The optimal values of b and τ, denoted b^* and τ^*, are related to a problem of optimally stopping a Markov process B called the 'risky fraction' process. The m dimensional process B is simply the vector of fractional allocations you would get if the initial allocation is B_0 and you never do any transactions; in other words, B_T^k would be the fraction of money held in stock k if no transactions ever occur.

The solution of the optimal stopping problem is fully described by a 'continuation region'

$$\mathscr{C} \subset \{b \in \mathbb{R}^m : b_1 > 0, \ldots, b_m > 0, b_1 + \ldots + b_m < 1\}.$$

It turns out that $\mathscr{C}$ is an open set which contains b^*. The optimal stopping rule τ^* for the portfolio manager is simply the first exit time from $\mathscr{C}$ by the process B which started with $B_0 = b^*$. In other words, having just rebalanced in order to achieve the allocation b^*, the portfolio manager does no transactions for a while but pays close attention to the fractional allocations among the m risky stocks. This is the same as watching the evolution of the risky fraction process B, which had been reset equal to b^* at the time of the last transaction. Moreover, τ^* will be the elapsed time between transactions. When B hits the boundary of $\mathscr{C}$, the next transactions are made, the portfolio is rebalanced to b^*, and the cycle is repeated.

To specify b^* and $\mathscr{C}$, it is necessary to specify the value function for the optimal stopping problem:

$$f_R(b) \equiv \sup_\tau \{-E_b[\ln(1-1'B_\tau)] - (R-r)E_b[\tau]\}.$$

Here R is a parameter $R > r$ whose role will be described below, and E_b denotes expectation conditioned on $B_0 = b$. Hence $f_R(b)$ is the value (that is, the expected net payoff) of being able to stop optimally the Markov process B when its initial value is b. A reward of $-\ln(1-1'B_\tau)$ is collected for stopping in state B_τ, but up until the time of stopping, a 'continuation fee' of $R-r$ per unit time is paid.

As shown by Morton & Pliska (1993), once you know the value function f, it is a simple matter to compute b^* and $\mathscr{C}$. The continuation region $\mathscr{C}$ is simply given by

$$\mathscr{C} = \{b \in \mathbb{R}^m : f_R(b) > -\ln(1-1'b)\}.$$

The optimal rebalance vector b^* will be the solution along with the parameter R of the $m+1$ equations:

$$1/(1-1'b) = \partial f_R(b)/\partial b^k, \quad k = 1, \ldots, m$$
$$0 = \ln\alpha + \ln(1-1'b) + f_R(b).$$

The value of the parameter R that is part of this solution will turn out to be the maximum growth rate for the portfolio.

Hence to solve this optimal portfolio problem, it suffices to compute the optimal stopping problem's value function $f_R(b)$. Morton & Pliska (1993) show this is given by the solution H of the following free boundary problem:

$$\tfrac{1}{2}\sum_i\sum_j H_{ij} b_i b_j[(e_i'-b')M(e_j-b)] + \sum_i H_i b_i[(e_i'-b')(\mu - r1 - Mb)] = R - r,$$
$$H(b) > h(b) \equiv -\ln(1-b_1-\ldots-b_m), \quad b \in \mathscr{C}^0,$$
$$H(b) = h(b), \quad b \in \partial\mathscr{C},$$
$$\partial H(b)/\partial b_i = \partial h(b)/\partial b_i, \quad i = 1, \ldots, m; \quad b \in \partial\mathscr{C}.$$

Here the subscripts on H denote partial derivatives, the subscripts on the vector b denote its components, and e_i denotes an m-component column vector consisting of zeros except for a 1 in the ith entry.

Morton & Pliska (1993) show that this free boundary problem is easy to solve numerically when there is only $m = 1$ risky stock, because then the boundary $\partial\mathscr{C}$ is fully characterized by two scalars. But the numerical methods required to solve this problem become much more complicated when $m \geqslant 2$, because then the boundary consists of infinitely many points. There are some standard approaches that can be considered, such as the discrete time, Markov chain approximation method of Kushner & Dupuis (1991), but their implementation is made extremely difficult by the non-constant coefficients in the partial differential equation.

This brings us to the purpose of this paper: to develop transformations of the above free boundary problem that lead to new free boundary problems which are easier to solve. In particular, we will focus on the case of $m = 2$ risky securities and show that the original free boundary problem can be transformed to one that is especially simple. In fact, Morton & Pliska (1993) use our transformations and Markov chain approximation methods to solve the problem numerically.

2. Main results

Our aim is to compute a solution of the free boundary problem when there are $m = 2$ risky securities. Equivalently, we want to compute the value function for the problem of optimally stopping the risky fraction process B when the reward-for-stopping function is $h(b)$ and there is a continuation cost-rate equal to $R-r$. It suffices, therefore, to specify three elements: (1) the partial differential equation; (2) the state space of the risky fraction process (a subset of which is the continuation region $\mathscr{C}$ which, in turn, is the domain for the solution of the partial differential equation); (3) the reward-for-stopping function. As we go through a series of transformations, these three elements will be identified at each stage. Hence each stage of this analysis will be associated with a unique free boundary problem as well as a unique optimal stopping problem.

In particular, our original problem will be described as follows: find a value function $H: \{b \in \mathbb{R}^2 : b_1 > 0, b_2 > 0,\ b_1 + b_2 < 1\} \to \mathbb{R}^2$ satisfying

$$\tfrac{1}{2} \sum_i \sum_j H_{ij} b_i b_j [(e_i' - b')\, M(e_j - b)] + \sum_i H_i b_i [(e_i' - b')(\mu - r1 - Mb)] = R - r \tag{1}$$

and corresponding to the reward-for-stopping function h.

In all that follows we denote $m_1 = m_{11}$, $m_2 = m_{22}$, and $m_3 = m_{12} = m_{21}$, where (m_{ij}) is the standard notation for the symmetric matrix $M = \Lambda\Lambda'$.

Proposition 1. *The function H satisfies* (1) *if and only if*

$$H(b_1, b_2) = \Phi(u_1, u_2),$$

where

$$u_i = b_i/(1 - b_1 - b_2), \quad i = 1, 2,$$

and where $\Phi: \mathbb{R}_+^2 \to \mathbb{R}^2$ is the value function satisfying

$$\tfrac{1}{2} m_1 u_1^2 \Phi_{11} + m_3 u_1 u_2 \Phi_{12} + \tfrac{1}{2} m_2 u_2^2 \Phi_{22} + (\mu_1 - r)\, u_1 \Phi_1 + (\mu_2 - r)\, u_2 \Phi_2 = R - r$$

and corresponding to the reward-for-stopping function

$$\phi(u_1, u_2) \equiv \ln(1 + u_1 + u_2).$$

Proof. A careful proof requires considerable tedious algebra; the presentation here will only sketch the main steps. First use the chain rule for partial derivatives to derive expressions for the five partial derivatives of H in terms of Φ. For example,

$$H_1 = \Phi_1 \frac{\partial u_1}{\partial b_1} + \Phi_2 \frac{\partial u_2}{\partial b_1} = \Phi_1 \frac{1 - b_2}{(1 - b_1 - b_2)^2} + \Phi_2 \frac{b_2}{(1 - b_1 - b_2)^2}.$$

Meanwhile, the partial differential equation in (1) can be written out as

$$\begin{aligned}
&\tfrac{1}{2}[m_1(1 - b_1)^2 - 2m_3(1 - b_1)\, b_2 + m_2 b_2^2]\, b_1^2 H_{11} \\
&+ [-m_1 b_1(1 - b_1) + m_3(1 - b_1)(1 - b_2) + m_3 b_1 b_2 - m_2 b_2(1 - b_2)]\, b_1 b_2 H_{12} \\
&+ \tfrac{1}{2}[m_1 b_1^2 - 2m_3 b_1(1 - b_2) + m_2(1 - b_2)^2]\, b_2^2 H_{22} \\
&+ [(\mu_1 - r)(1 - b_1) - m_1 b_1(1 - b_1) - m_3(1 - b_1)\, b_2]\, b_1 H_1 \\
&- [(\mu_2 - r)\, b_2 - m_3 b_1 b_2 - m_2 b_2^2]\, b_1 H_1 \\
&- [(\mu_1 - r)\, b_1 - m_1 b_1^2 - m_3 b_1 b_2]\, b_2 H_2 \\
&+ [(\mu_2 - r)(1 - b_2) - m_3 b_1(1 - b_2) - m_2 b_2(1 - b_2)]\, b_2 H_2 = R - r.
\end{aligned} \tag{2}$$

Substituting the expressions for the partial derivatives H_1, H_2, H_{11}, H_{12}, and H_{22} into (2), wading through a lot of tedious algebra, and eventually switching to the new variables u_1 and u_2, one finally obtains the partial differential equation in the hypothesis of this proposition. The new reward-for-stopping function ϕ is immediately obtained with the same change of variables. □

The first transformation has greatly simplified the coefficients in the partial differential equation, although they are still not constants. The next transformation will lead to a partial differential equation with constant coefficients.

Proposition 2. *The function Φ satisfies the problem in Proposition* 1 *if and only if*

$$\Phi(u_1, u_2) = \Delta(z_1, z_2),$$

where $$z_i = \ln(u_i)/\surd m_i, \quad i = 1, 2,$$

and where $\Delta : \mathbb{R}^2 \to \mathbb{R}^2$ *is the value function satisfying*

$$\tfrac{1}{2}\Delta_{11} + \frac{m_3}{\surd(m_1 m_2)}\Delta_{12} + \tfrac{1}{2}\Delta_{22} + \frac{1}{\surd m_1}(\mu_1 - r - \tfrac{1}{2}m_1)\,\Delta_1 + \frac{1}{\surd m_2}(\mu_2 - r - \tfrac{1}{2}m_2)\,\Delta_2 = R - r$$

and corresponding to the reward-for-stopping function

$$\delta(z_1, z_2) \equiv \ln\{1 + \exp[\surd(m_1)\, z_1] + \exp[\surd(m_2)\, z_2]\}.$$

The proof will be omitted, because it can be carried out in the same manner as with Proposition 1. Looking at the partial differential equation in Proposition 2, one sees that it corresponds to a two-dimensional brownian motion with state space all of $\mathbb{R}^2$, with drift, and with correlated components. The next transformation will eliminate the cross-partial term.

Proposition 3. *The function* Δ *satisfies the problem in Proposition 2 if and only if*

$$\Delta(z_1, z_2) = \Theta(x_1, x_2),$$

where $$x_1 = cz_1 + dz_2, \quad x_2 = dz_1 + cz_2, \quad \nu \equiv m_3/\surd(m_1 m_2),$$

$$c \equiv -\nu/\surd\{2(1-\nu^2)[1 + \surd(1-\nu^2)]\} \quad \textit{and} \quad d \equiv \surd\{[1 + \surd(1-\nu^2)]/2(1-\nu^2)\},$$

and where $\Theta : \mathbb{R}^2 \to \mathbb{R}^2$ *is the value function satisfying*

$$\tfrac{1}{2}\Theta_{11} + \tfrac{1}{2}\Theta_{22} + a\Theta_1 + b\Theta_2 = R - r$$

and corresponding to the reward-for-stopping function

$$\theta(x_1, x_2) \equiv \ln\{1 + \exp[(px_2 - qx_1)\surd m_1] + \exp[(px_1 - qx_2)\surd m_2]\},$$

where $$p \equiv d/(d^2 - c^2), \quad q \equiv c/(d^2 - c^2),$$

$$a \equiv \frac{c}{\surd m_1}(\mu_1 - r - \tfrac{1}{2}m_1) + \frac{d}{\surd m_2}(\mu_2 - r - \tfrac{1}{2}m_2),$$

and $$b \equiv \frac{d}{\surd m_1}(\mu_1 - r - \tfrac{1}{2}m_1) + \frac{c}{\surd m_2}(\mu_2 - r - \tfrac{1}{2}m_2).$$

Note that M is positive definite, so $\nu^2 < 1$; it follows that c and d are real with $d^2 > c^2$. Hence all the quantities in Proposition 3 are well defined.

Proof. Substituting expressions for the partial derivatives as in the proof of Proposition 1, one eventually obtains

$$\tfrac{1}{2}(c^2 + 2\nu cd + d^2)\,\Theta_{11} + (\nu c^2 + 2cd + \nu d^2)\,\Theta_{12} + \tfrac{1}{2}(\chi^2 + 2\nu cd + d^2)\,\Theta_{22} + a\Theta_1 + b\Theta_2 = R - r.$$

A little more algebra reveals that $c^2 + 2\nu cd + d^2 = 1$ and that $\nu c^2 + 2cd + \nu d^2 = 0$, so we actually have the partial differential equation as stated in the hypothesis of this proposition.

With regard to the reward-for-stopping function, notice that $d^2 > c^2$ implies the change of variables is non-singular. It is easily shown, therefore, that

$$z_1 = px_2 - qx_1 \quad \text{and} \quad z_2 = px_1 - qx_2.$$

Hence the reward-for-stopping function θ is as indicated. □

We now have the problem of stopping a two-dimensional brownian motion on $\mathbb{R}^2$, where the reward-for-stopping function is θ and the continuation rate is $R-r$. The brownian motion components are independent of each other, but they each have a non-zero drift. This is the problem that Morton & Pliska (1993) solved by using a Markov chain approximation. By taking the transformations in the reverse direction, they were then able to estimate the original value function H. For example, having estimated the value of Θ at the point (x_1, x_2), they knew that this equalled the value of Δ at the point $z_1 = px_2 - qx_1, z_2 = px_1 - qx_2$, the value of Φ at the point where $u_i = \exp[\surd(m_i) z_i], i = 1, 2,$ and finally the value of H at the point where $b_i = u_i/(1+u_1+u_2), i = 1, 2.$

3. Summary and conclusions

In this paper we focus on a non-constant coefficients, two-dimensional partial differential equation which arises in the optimal management of a portfolio when transaction costs are fixed. Non-constant coefficient, multidimensional, second-order partial differential equations arise naturally in finance. The degree of the partial differential equation reflects the modelling of uncertainty. Cross-product terms arise because of correlations between asset prices and/or state variables. The simplest model that we use for describing an asset price is geometric brownian motion. This leads to equations with unbounded coefficients and the vanishing of leading terms (see, for example, Gleit 1978). There is an increasing use of two- and three-dimensional equations, particularly in the fields of options on many assets and stochastic volatility. In addition, nonlinear equations are also arising out of foundational issues (see, for example, He & Leland 1993; Hodges & Carverhill 1993). Therefore, many of the standard approaches for analysis cannot be used without considerable care. Further, many standard numerical analysis schemes are not robust with respect to the partial differential equations arising in finance (see Clewlow 1990).

A major feature of these types of problems is that a direct attack may lead to extreme difficulties. However, we believe that, whenever possible, suitable transformations of the equations should be undertaken to make them more analytically and computationally tractable. The determination of the transformations and their implementation are of fundamental importance. Although we have not used Lie groups directly in this paper, we believe that Lie groups and normal form theory may be exactly that part of mathematics which will make a crucial contribution to the continuation of the revolution instigated by the introduction of martingales and stochastic integrals in the theory of continuous trading.

In the paper we have not derived the normal form for the original partial differential equation. The normal form reduces to a Schrödinger equation. We have not worked with the Schrödinger equation form because we are able, in this case, to maintain our insight in regard to the brownian motions and the use of the Markov chain approximation. However, we believe that it is quite likely that if we had, say, worked with asset prices, where the instantaneous means and standard deviation of returns were not constant, then a final reduction to the normal form would have been necessary to give meaningful insights and to provide computational advantages. In particular, we may have been led to computational advantages based on perturbation or asymptotic analysis (provided they could have been validly carried out). Clearly we are suggesting this as a further area for research. Indeed, Atkinson & Wilmott (1993) have already commenced work in this field by explicitly attacking the original Morton & Pliska (1993) model with an asymptotic analysis.

References

Atkinson, C. & Wilmott, P. 1993 Portfolio management with transaction costs: an asymptotic analysis. Mathematics Department, Imperial College. (Working paper.)

Clewlow, L. J. 1990 Finite difference techniques for one and two dimension option valuation problems. FORC Preprint 90/10, University of Warwick.

Gleit, A. 1978 Valuation of general contingent claims: existence, uniqueness, and comparisons of solutions. *J. Financial Econ.* **6**, 71–87.

He, H. & Leland, H. 1993 On equilibrium asset price processes. *Rev. Financial Stud.* **6**, 593–617.

Hodges, S. & Carverhill, A. 1993 Quasi mean reversion in an efficient stock market: the characterization of economic equilibria which support Black–Scholes option pricing. *Economic J.* **103**, 395–405.

Kushner, H. & Dupuis, P. 1991 *Numerical methods for stochastic control problems in continuous time.* New York: Springer-Verlag.

Morton, A. & Pliska, S. 1993 Optimal portfolio management with fixed transaction costs. *Math. Finance.* (In the press.)

11

Interest rate volatility and the shape of the term structure

Roger H. Brown and Stephen M. Schaefer

This paper analyses the effect of interest rate uncertainty on the shape of the forward rate curve. We consider a broad class of term structure models characterized by an affine relation between the drift and diffusion coefficients of the stochastic process describing the evolution of the state variables and the level of the state variables. For these models, a simple relation exists between the shape of the forward rate curve, the sensitivity of the zero-coupon yield curve to the state variables and the variance–covariance matrix of the state variables. In single factor models this relation implies that minus the convexity of the forward rate curve with respect to a measure of 'duration' is equal to the variance of the short rate.

The paper explores why it is that, despite the well known shortcomings of single factor models, attempts to fit such models to cross-sections of nominal bond prices nonetheless produce reasonable estimates of interest rate volatility.

1. Introduction

Interest rate volatility plays a central role in modern term structure theory and, in particular, causes the forward rate curve and the expected path of short rates to diverge. In contrast, classical theory predicts that in the absence of risk premia, forward rates will be equal to expected short rates. The modern theory shows clearly that this will not generally be the case, even when risk premia are zero, unless the path of future interest rates is certain.

In the modern theory, the price of a zero-coupon bond may be written as the expectation of the value of the bond's pay-off, discounted at future short rates. The expectation is taken under the appropriate 'risk adjusted' probability measure and, in this case, the bond price may be written as

$$P(r,t) = \hat{E}_t\left(\exp - \int_t^T r(s)\,\mathrm{d}s\right),$$

where $\{r(s),\ s \in [t, T]\}$ is the path of future short rates between the current time, t, and the maturity date of the bond, T. The operator $\hat{E}_t$ denotes an expectation at time t taken with respect to the risk adjusted probability measure. (In the continuous time theory of contingent claims valuation, an 'economic' argument – such as no-arbitrage or equilibrium – is used to derive a partial differential equation known as the 'valuation equation'; see, (2) for example. The 'risk-adjusted' process is an Îto process in which the drift and diffusion coefficients are set equal to the corresponding

Mathematical Models in Finance. Edited by S.D. Howison, F.P. Kelly and P. Wilmott.
Published in 1995 by Chapman & Hall, London. ISBN 0 412 63070 2

coefficients of the valuation equation.) As exp ($\cdot$) is a convex function we have, from Jensen's inequality,

$$P(r,t) = \hat{E}_t\left(\exp - \int_t^T r(s)\,\mathrm{d}s\right) > \exp\left(\hat{E}_t - \int_t^T r(s)\,\mathrm{d}s\right),$$

or, in other words, the bond price is higher under the modern theory than if the forward rate curve were equal to the (risk adjusted) path of the expected short rates. This 'price premium', or 'yield discount', is a direct consequence of uncertainty in future interest rates and is largely unnoticed in traditional theory.

If interest rate uncertainty is central to the modern theory, what is the main feature of the term structure in which interest rate uncertainty is reflected? This is the main question we try to answer in this paper.

We also attempt to throw some light on a puzzle which has emerged in recent applications of modern term structure theory to bond market data. Brown & Dybvig (1986) (BD) have analysed US Treasury bills, notes and bonds; Barone *et al.* (1989) (BCZ) have analysed Italian Treasury bonds and, in previous papers, Brown & Schaefer (1988, 1994) (BS) have analysed UK Government Index-Linked bonds. All three studies fit the (single factor) Cox *et al.* (1985) (CIR) model to cross-sectional data.

The results of these studies are interesting here because, whereas some predictions of the model are strongly rejected by the data, others, and particularly the results relating to interest rate volatility implied from the yield curve, appear to fit the data quite well. For example, both BD and BCZ (using nominal bond data) find that the implied long rate of interest, a constant in the CIR model, is highly variable. BS (1994) find that parameter stability can be firmly rejected (although the implied long term *real rate* shows a much higher degree of stability). In these respects the results are fundamentally inconsistent with the model. On the other hand all three studies find a relatively high degree of correspondence between interest rate volatility implied by cross-sections of bond prices, i.e. from the shape of the yield curve, and the volatility observed in time series. Why is this feature of the model supported by the data whereas others are clearly rejected? We show below that there is a relatively simple relation between interest rate volatility and the shape of the yield curve which is shared by a broad class of models including the CIR model. Thus success in imputing interest rate volatility from the yield curve does not imply support for the CIR model alone.

In a recent paper Litterman *et al.* (1991) (LSW) investigate the relation between the shape of the US Treasury yield curve and the volatility of yield changes implied by the prices of options on Treasury bond futures. Using four and a half years of weekly data, they find that a linear function of the one month, three-year and ten-year yields explains around 70% of the variability in implied volatility.

The remainder of the paper is organized as follows. In §2 we define a class of time homogeneous single factor term structure models (the *affine yield class*) for which a relatively simple relation exists between the forward rate curve, a measure of 'duration' and the level of interest rate volatility. Section 3 provides a complete characterization of this class and §4 discusses multivariate affine models. In §5 we ask what can be deduced from the empirical literature about the factor structure of yield curve movements and about the sensitivity of the yield curve to factor shifts. This allows us, in §6, to extend our analysis to a two-factor affine model which, we suggest, may help to explain the puzzle described above. Section 7 gives our conclusions.

2. The affine yield class of term structure models

We consider the class of time homogeneous single factor models. (We use the term 'time homogeneous' to refer to the case where the drift and diffusion coefficients of the valuation equation, e.g. (2), are, at most, functions of the level of the state variable and do not explicitly depend on time. Such models have the property that the yield curve at any point in time depends only on the state variable, e.g. the level of the short term rate of interest, and not on calendar time. Throughout this paper we restrict our attention to time homogeneous models.) We take the state variable to be the instantaneous short rate, r, which is assumed to follow a diffusion process with drift $\mu(r)$ and variance $\sigma^2(r)$,

$$\mathrm{d}r = \mu(r)\,\mathrm{d}t + \sigma(r)\,\mathrm{d}z. \tag{1}$$

Under appropriate conditions (see, for example, CIR (1985)) the value, $P(r,t)$, of any interest rate contingent claim must satisfy

$$\alpha(r)\,P_r + P_t + \tfrac{1}{2}\sigma^2(r)\,P_{rr} - rP = 0, \tag{2}$$

where $\alpha(r)$ is the so-called 'risk adjusted' drift of the short rate process and is equal to $\mu(r)$ plus a term which incorporates the market risk premium (see, for example, CIR (1985), p. 393).

The price at time t, $P(r,t,t+\tau)$, of a zero-coupon bond maturing at time $t+\tau$, and with current time-to-maturity τ, is obtained by solving (2) subject to the terminal condition,

$$P(r,t+\tau,t+\tau) = 1. \tag{3}$$

In a number of cases the solution for a zero-coupon bond has a particularly simple form with the price being given by

$$P(r,t,t+\tau) = \mathrm{e}^{A(\tau)-B(\tau)\,r(t)}, \tag{4}$$

where $A(\tau)$ and $B(\tau)$ are functions of time-to-maturity, τ. (In writing $A(\cdot)$ and $B(\cdot)$ simply as functions of time-to-maturity, τ, it is important to remember that $\tau = T-t$, where T is the maturity date and t is the current date. Thus, for a given bond (with a fixed maturity date, T) time-to-maturity, τ, is a function of t.) From (4) the corresponding zero-coupon yield, $R(r,t,t+\tau)$, is given by

$$R(r,t,t+\tau) = -(1/\tau)\ln[P(r,t,t+\tau)] = -[A(\tau)/\tau] + [B(\tau)/\tau]\,r(t). \tag{5}$$

Thus, in these cases, the zero-coupon yield is *affine* in the short rate, r, and we shall refer to models that possess this property as belonging to the *affine yield class*.

Definition (*affine yield class*). A single factor term structure model belongs to the affine yield class if zero-coupon yields under the model are affine in the short rate.

A number of well known models belong to this class (for example, Vasicek 1977; CIR 1985). We discuss these and a number of other cases below.

For models belonging to the affine yield class and which are also time homogeneous it is possible to derive a simple relation between the forward rate of interest, the parameters of the risk adjusted process and the function $B(\tau)$.

Proposition 1. *For models that belong to the affine yield class and are time homogeneous, the instantaneous forward rate, $f(r, t, t+\tau)$, at time t for the instantaneous future period at date $t+\tau$ is given by*

$$f(r,t,t+\tau) = r(t) + \alpha(r)B(\tau) - \tfrac{1}{2}\sigma^2(r)B^2(\tau). \tag{6}$$

Proof. From (4), we have

$$P_r/P = -B(\tau) \quad \text{and} \quad P_{rr}/P = B^2(\tau), \tag{7}$$

and, we also have

$$P_t/P = f(r,t,t+\tau). \tag{8}$$

(Although some recent time inhomogeneous models also belong to the affine yield class, the same simple characterization in terms of forward rates does not apply because (8) applies only in the time homogeneous case.) Substituting (7) and (8) into (2) gives (6). □

The function $B(\tau)$ which, as (6) shows, largely determines the shape of the term structure, has two (related) interpretations. First, $B(\tau)$ determines the term structure of yield volatility. From (5) the volatility of the τ-period zero-coupon yield $R(r,t,t+\tau)$ is given by

$$[B(\tau)/\tau]\,\sigma(r). \tag{9}$$

Second, as CIR (1979) have suggested, $B(\tau)$ may be regarded as a duration measure because, like standard duration, it is equal to minus the semi-elasticity of the bond price with respect to an interest rate, in this case, the short rate (see (7)). Equation (6) shows that the forward rate is a quadratic (concave) function of the duration measure, $B(\tau)$. The volatility parameter, $\sigma^2(r)$, affects the forward rate curve in two ways. First, for some, but not all members of the affine yield class, the function $B(\tau)$ is itself a function of $\sigma^2(r)$. Second, and more directly, changes in volatility affect the convexity of the forward rate curve in interest rate duration, i.e. $B(\tau)$, space. Indeed the convexity of the forward rate curve with respect to $B(\tau)$ is simply $-\sigma^2(r)$. In other words, we have

Proposition 2. *For models that belong to the affine yield class and are time homogeneous, minus the second derivative of the forward rate of interest, $f(r,t,t+\tau)$, with respect to the 'duration' function $B(\tau)$ is equal to the instantaneous variance of the short rate*

$$-\partial^2 f(r,t,t+\tau)/\partial B^2(\tau) = \sigma^2(r).$$ □

This relation provides not only intuition but also a route for empirical investigation.

3. Characterizing the affine yield class

The simple relation between the term structure of forward rates and the term structure of volatility (or 'duration') applies only to those models which belong to the time homogeneous affine yield class. In this section we give the conditions on $\alpha(r)$ and $\sigma^2(r)$ which are necessary and sufficient for a model to belong to the class. First, we show that, when a model belongs to the class, $\alpha(r)$ is affine in r if and only if $\sigma^2(r)$ is affine in r.

Proposition 3. *For a single factor model which belongs to the time homogeneous affine yield class, the risk adjusted drift, $\alpha(r)$ is affine if and only if the risk adjusted variance, $\sigma^2(r)$ is affine.*

Proof. Direct substitution of (4) and (7) into (2) gives

$$-B(\tau)\,\alpha(r) = A_\tau(\tau) - B_\tau(\tau)\,r - \tfrac{1}{2}\sigma^2(r)B^2(\tau) + r, \tag{10}$$

where a τ subscript denotes the partial derivative with respect to τ.

Now, suppose $\sigma^2(r)$ is affine, then, by inspection, $\alpha(r)$ is affine. Similarly, when $\alpha(r)$ is affine, $\sigma^2(r)$ is affine. Hence $\alpha(r)$ is affine if and only if $\sigma^2(r)$ is affine. □

Next we show that, when both $\alpha(r)$ and $\sigma^2(r)$ are time homogeneous, a model belongs to the time homogeneous affine yield class if and only if $\alpha(r)$ and $\sigma^2(r)$ are affine in r.

Proposition 4. *If both $\alpha(r)$ and $\sigma^2(r)$ are time homogeneous, a model belongs to the time homogeneous affine yield class if and only if both $\alpha(r)$ and $\sigma^2(r)$ are affine in r.*

Proof. If (4) solves the fundamental valuation (2) in the case of a zero-coupon bond, i.e. subject to the boundary condition (3), and $\alpha(r,t)$ and $\sigma^2(r,t)$ are time homogeneous, then both $\alpha(r)$ and $\sigma^2(r)$ are affine in r. Re-arranging (10), we have

$$\alpha(r) = -(A_\tau/B) - [(1-B_\tau)/B]\,r + \tfrac{1}{2}B\sigma^2(r). \tag{11}$$

Differentiating with respect to r gives

$$\partial\alpha(r)/\partial r = -(1-B_\tau)/B + \tfrac{1}{2}B\,\partial\sigma^2(r)/\partial r. \tag{12}$$

Remembering that $\alpha(\cdot)$ and $\sigma^2(\cdot)$ are functions of r only and differentiating with respect to t gives

$$0 = \frac{\partial^2\alpha(r)}{\partial r\,\partial t} = -\left[\frac{BB_{\tau\tau}+B_\tau-B_\tau^2}{B^2} + \tfrac{1}{2}B_\tau\frac{\partial\sigma^2(r)}{\partial r}\right]. \tag{13}$$

(In deriving (13) we remember that, for a given maturity date, T, time-to-maturity, τ, is equal to $T-t$. It follows that $B_t = -B_\tau$, etc. See note in parentheses after (4).) From (13) we obtain

$$\frac{\partial\sigma^2(r)}{\partial r} = -\frac{2}{B_\tau}\frac{(BB_{\tau\tau}+B_\tau-B_\tau^2)}{B^2}. \tag{14}$$

The right-hand side of (14) is at most a function of time and, since $\sigma^2(r)$ is not a function of time, we have that

$$\partial\sigma^2(r)/\partial r = \text{const}, \tag{15}$$

i.e. that $\sigma^2(r)$ is affine in r. Further, from Proposition 3, it follows that, if $\sigma^2(r)$ is affine in r, then $\alpha(r)$ is affine in r.

Finally we must show that when $\alpha(r)$ and $\sigma^2(r)$ are time homogeneous and affine in r then (4) solves (2) subject to (3). Let

$$\alpha(r) = a_0 + a_1 r, \tag{16}$$

and

$$\sigma^2(r) = b_0 + b_1 r. \tag{17}$$

To complete the proof it is sufficient to show that, for each combination of zero and non-zero coefficients in (16) and (17), there is a corresponding $A(\tau)$ and $B(\tau)$ for which (4) solves (2). In fact each relevant combination has already been considered in the literature. Thus, for example, when a_0, a_1 and b_0 are non-zero and b_1 is zero we have Vasicek's model and when a_0, a_1 and b_1 are non-zero and b_0 is zero we have the CIR model. Other cases include Merton (1971), for which a_0, a_1 and b_1 are zero and only b_0 is non-zero and Pearson & Sun (1988) for which all four parameters are non-zero. In each of these cases (4) solves (2) and it follows that, in general, when $\alpha(\cdot)$ and $\sigma^2(\cdot)$ are time homogeneous and affine in r, (4) solves (2). □

The form of $B(\tau)$ depends on a_1 and b_1, the coefficients of r in (16) and (17). When a_1 is non-zero, say $-\kappa$, and b_1 is equal to zero we have Vasicek's (1977) model and the solution for $B(\tau)$, subject to the appropriate boundary condition, is

$$B(\tau) = (1-e^{-\kappa\tau})/\kappa. \tag{18}$$

When both a_1 and b_1 are non-zero ($a_1 = -(\kappa+\lambda)$ and $b_1 = \sigma^2$, say) we have the CIR model and $B(\cdot)$ is given by

$$B(\tau) = 2(e^{\gamma\tau}-1)/[(\gamma+\kappa+\lambda)(e^{\gamma\tau}-1)+2\gamma], \tag{19}$$

where

$$\gamma = \surd[(\kappa+\lambda)^2+2\sigma^2]. \tag{20}$$

4. Multivariate affine yield models

Duffie & Kan (1993) (DK) have recently extended Proposition 4 to the multivariate case. They consider a model with n factors, $X_1(t), X_2(t) \ldots X_n(t)$, such that, for all t and τ, the price of a zero-coupon bond, $P(t, \tau)$ is a function $P(t, \tau, X)$. Letting μ denote the drift in X, $\sigma\sigma^{\mathrm{T}}$ the covariance matrix of X, and $C(\tau)$ a matrix which depends on the maturity dependent coefficients (i.e. those coefficients which do not depend on X) in the pricing solution, DK prove the following.

Proposition 5. *Suppose* $(P, \mu, \sigma\sigma^{\mathrm{T}})$ *is a compatible term structure model. If* μ, $\sigma\sigma^{\mathrm{T}}$ *and* r *are affine then* P *is exponential affine. Conversely, if* P *is exponential affine and there exist maturities* $t_1, t_2, \ldots, t_n$ *such that* $C(t_1, t_2, \ldots, t_n)$ *is non-singular, then* μ, $\sigma\sigma^{\mathrm{T}}$ *and* r *are affine. (Duffie & Kan define 'compatible' in this context to mean a choice of* P, μ *and* σ *such that*

$$P(X, T-t) = \hat{E}_t\left[\exp\left(-\int_t^T r(s)\,\mathrm{d}s\right)|X_t\right]\text{a.s.,}\quad 0 \leqslant t \leqslant T < \infty.)$$

In fact both Propositions 4 and 5 are to some extent anticipated in earlier results by CIR. CIR (1980) show that in a single factor model where the drift in the short rate is affine in r, long rates are affine in r if and only if both $\sigma^2(r)$ and the market risk adjustment term in the drift are also affine in r. This result is similar to Proposition 4 except that CIR assume a particular (affine) form for the drift whereas Proposition 4 shows that the affine solution arises if and only if both the drift and the diffusion terms are affine in r.

CIR (1981) consider a multivariate model where the state variables are the short rate, r, and exponentially smoothed averages of r. Assuming the drift in r is affine in r, CIR show that the exponential affine pricing relation results if and only if both $\sigma^2(r)$ and the market risk adjustment term are affine in r. Again, this is similar to Proposition 5 but Duffie & Kan's result is more general in two respects. First, as in the single-factor case, CIR do not consider necessary and sufficient conditions on the drift term. Second, in the CIR model the state variables other than the short rate are all locally deterministic; no such restriction applies in the DK model. (A number of multivariate affine yield models exist in the literature, e.g. CIR (1985) and Longstaff & Schwartz (1991).)

5. Empirical evidence on interest rate dynamics

(*a*) *The factor structure*

Multivariate theoretical models such as DK are capable of accommodating a wide variety of factor structures and dynamics. However, not all these possibilities are of empirical importance and recent empirical work has produced some interestingly consistent evidence on the nature of yield curve dynamics. Using these results and our earlier analysis, §6 resurrects a version of Proposition 2 which holds even in the multivariate case.

Using principal components analysis and data on the US Treasury market over the period January 1984 to June 1988, Litterman & Scheinkman (1991) (LS) find that two factors explain around 98% of changes in estimated zero-coupon yields and about 97% of bond returns. The first principal component explains on average 90% of the variability of yield curve movements but only around 80% of the variability of short rates. Caverhill & Strickland (1992) (CS) obtain very similar results using UK and US swap rate data over the period April 1989 to August 1990, as do Nelson & Schaefer (1983) (NS) using data on US Treasuries for the period 1930 to 1979. (Both Litterman & Scheinkman and Caverhill & Strickland find a third factor which accounts for almost all the remaining variation in yields. This factor has a convex pattern of loadings over yields which, following the analysis of §2, suggests that the third factor is connected with changes in anticipated volatility.) Thus it appears that though one factor is insufficient to characterize the covariance structure of nominal yield changes, we do not need to think in terms of five or ten factors but rather two or possibly, three. In what follows we focus on the two factor case.

There is also a high degree of consistency in the sensitivity of the zero-coupon yield curve to the two most important factors. Both LS and CS find that the sensitivity of zero-coupon yields to the first principal component (also referred to as the 'yield curve impact' of the relevant principal component) is close to unity for all maturities. This is also consistent with the results obtained by NS, who pre-specified the factors as a long term interest rate and the difference (or spread) between the long and short term rates. NS found the sensitivity of yields to the long rate to be close to unity.

In the remainder of the paper we assume, following Schaefer & Schwartz (1984) that the two factors determining the term structure are the long rate, l, and the spread, S, defined as $(r-l)$, the difference between the short rate, r, and the long rate, l. Evidence on the yield curve impact of the spread is less clear from the literature but, as we shall demonstrate, there is no doubt that it shows considerably more attenuation with maturity than the impact of the long rate. However, to compare the various estimates of yield sensitivity we must first adjust for the different definitions of the 'long rate' and the 'short rate' (and consequently, of 'the spread') which different studies have used.

Assume a two-factor model where, for some definition of l and S, the change in the τ_j-period zero-coupon yield is given by

$$\Delta R(\tau_j) = \alpha_{0j} + \alpha_{sj}\,\Delta S + \alpha_{lj}\,\Delta l. \tag{21}$$

Now assume we attempt to estimate (21) but that we define

$$l^* \equiv R(\tau_b) \quad \text{and} \quad S^* \equiv R(\tau_a) - R(\tau_b).$$

We may now write the change in any zero-coupon yield as

$$\Delta R(\tau_j) = \alpha^*_{0j} + \alpha^*_{sj}\,\Delta S^* + \alpha^*_{lj}\,\Delta l^*. \tag{22}$$

Assuming that α_{lj}, the sensitivity to the long rate, is unity for all maturities, it is straightforward to show that α^*_{sj}, the sensitivity of $R(\tau_j)$ to the new definition of the spread, S^*, is given by

$$\alpha^*_{sj} = (\alpha_{sj} - \alpha_{sb})/(\alpha_{sa} - \alpha_{sb}). \tag{23}$$

Using a simple equilibrium model we may characterize α_{sj} in terms of the rate of mean reversion in the spread. Using this relation and (23) we may then estimate the implied rate of mean reversion in the spread from empirical estimates of α^*_{sj}.

(b) Estimating the rate of mean reversion in the spread

For simplicity we assume that both l and S are gaussian, uncorrelated (Nelson & Schaefer, following Ayres & Barry (1979, 1980), find that the correlation between the long rate and the spread is typically quite small) and that their evolution is described by the risk adjusted processes

$$dl = \kappa_l(\mu_l - l)\,dt + \sigma_l\,dz_l, \tag{24}$$

$$dS = \kappa_s(\mu_s - S)\,dt + \sigma_s\,dz_s, \tag{25}$$

where κ_l and κ_s are the risk adjusted rates of mean reversion, μ_l and μ_s the risk adjusted long run means and σ_l and σ_S the local standard deviations of the long rate and spread. Following our earlier analysis, the forward rate at time t for the instantaneous future period at time $t+\tau$ is given by

$$f(r, t, t+\tau) = l + S + \alpha(S)\,B_s(\tau) - \tfrac{1}{2}\sigma_s^2 B_s^2(\tau) + \alpha(l)\,B_l(\tau) - \tfrac{1}{2}\sigma_l^2 B_l^2(\tau), \tag{26}$$

where $\alpha(S) \equiv \kappa_s(\mu_s - S)$, $\alpha(l) \equiv \kappa_l(\mu_l - l)$ and

$$B_s(\tau) = (1 - e^{-\kappa_s \tau})/\kappa_s \quad \text{and} \quad B_l(\tau) = (1 - e^{-\kappa_l \tau})/\kappa_l. \tag{27}$$

From (26) it follows that α_{sj}, the sensitivity of the τ_j-period zero-coupon yield to the spread, is given by:

$$\alpha_{sj} = B_s(\tau_j)/\tau_j = (1 - e^{-\kappa_s \tau_j})/\kappa_s \tau_j. \tag{28}$$

Substituting (28) into (23) we obtain an expression for the sensitivity of the τ_j-period zero-coupon yield to the spread, for any definition of the spread.

We have estimated sensitivities for six major currencies using weekly interest rate swap data for various periods between 1987 and 1993. These estimates were obtained from the following regression

$$\Delta R(\tau_j) = \alpha_{0j}^* + \alpha_{sj}^*\,\Delta S^* + \alpha_{lj}^*\,\Delta l^* + \tilde{u}, \tag{29}$$

where $\Delta R(\tau_j)$ is the change in the j-period zero-coupon yield, l^* is defined as the 10-year zero-coupon rate and S^* as the difference between l^* and the two-year rate. Figure 1 shows the estimated sensitivities to (*a*) the long rate and (*b*) the spread. The figure shows that the results are broadly consistent across the six currencies with the long rate sensitivities approximately unity for all maturities and the spread sensitivity attenuating quite quickly with maturity.

Estimates of κ_s and κ_l were obtained by fitting (23) and (28) to the estimated sensitivities using nonlinear least squares. The results are shown in table 1 which also includes estimates of κ_s and κ_l from yield curve sensitivities derived from direct estimates of zero-coupon yields in the US and UK government bond markets. (The direct estimates of the US and UK zero-coupon yield curves were obtained using the tax specific approach described in Schaefer (1981) for a tax exempt clientele.) Once again the results are quite consistent. The estimates of κ_s are generally in the range 0.3 to 0.7, although the estimate for the Italian Lire swap curve is appreciably higher at 1.51. The estimates of κ_l, on the other hand, are all close to zero with a mean of 0.0026. (As κ_l tends to zero, $B_l(\tau)$ tends to τ and the corresponding sensitivity to the long rate tends to unity for all maturities. A value for κ_l of 0.0026 implies that the sensitivities, which are unity at τ = zero and monotonic decreasing, are still 0.997 at 30 years.)

Thus the empirical evidence on the dynamics of the yield curve shows a substantial

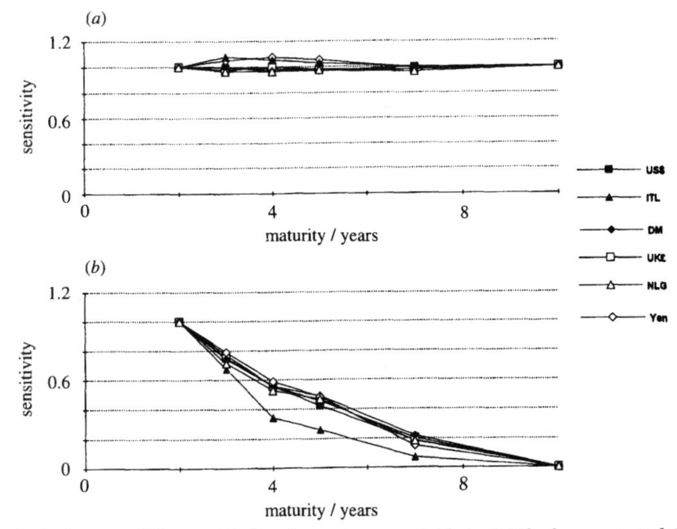

Figure 1. Estimates of the sensitivity of zero-coupon yields to (*a*) the long rate and (*b*) the spread derived from estimating (29). Data used are swap rates for maturities between 2 years and 10 years. In the regression the 'long rate' is defined as the 10-year rate and the spread as the difference between the 10-year and 2-year rates.

Table 1. *Implied mean reversion parameters from yield curve sensitivities*
(Estimates of κ_S and κ_l from fitting (23) and (28) to estimates of zero-coupon yield sensitivities to the long rate and the spread.)

currency	data source	frequency	period	κ_S	κ_l
US$	direct term structure estimates from Treasury bonds[a]	monthly	1930–1979	0.778	0.0225
UK£	direct term structure estimates from Gilts	monthly	1984–1989:03	0.219	−0.0071
NLG	swap rates (2–10 yrs)	weekly	1991:06–1993:03	0.443	0.0064
US$	swap rates (2–10 yrs)	weekly	1987:04–1993:06	0.418	0.0024
UK£	swap rates (2–10 yrs)	weekly	1987:04–1993:06	0.389	0.0030
Yen	swap rates (2–10 yrs)	weekly	1989:09–1993:07	0.347	−0.0064
DM	swap rates (2–10 yrs)	weekly	1989:09–1993:07	0.352	0.0056
ITL	swap rates (2–10 yrs)	weekly	1990:11–1993:06	1.512	−0.0057

[a] Source: Nelson & Schaefer (1983).

degree of consistency. Two main factors are identified which we take to be the long rate and the spread. The sensitivity of zero-coupon yields to changes in the long rate is close to unity for all maturities, implying a mean reversion coefficient which is close to zero. The sensitivity of zero-coupon yields to the spread attenuates relatively quickly with maturity and most implied mean reversion coefficients are in the range 0.3 to 0.7 with an average of 0.56. As we shall see, the much lower rate of attenuation in the sensitivity to the long rate implies that it is largely the variance of the long rate, rather than the spread variance, which determines the curvature of the yield curve other than for short maturities.

6. The shape of the term structure in a bivariate affine model

(*a*) *Curvature in the term structure for long maturities*

We have shown earlier that, in single factor affine models, the variance of the short rate is equal to minus the convexity of the forward rate curve measured with respect to the 'duration' measure defined in (4). In models with more than one state variable, such an interpretation is no longer possible, not least because there is now one duration measure and one variance associated with each state variable. However, making empirically realistic assumptions about the number of factors and the duration measures, we may show that a strong relation still exists between the variance of one of the factors and the shape of a part of the term structure. We suggest that this relation may help to explain a puzzle in the empirical literature.

Following the analysis of §5, we again consider the two-factor model based on the long rate and the spread and described in (24), (25) and (26). We also make use of the fact that B_s and B_l have the same functional form; only the mean reversion coefficients differ. We therefore compute the convexity of the forward curve with respect to a duration measure B_x which has the same form as B_s and B_l but has a mean reversion coefficient κ_x

$$B_x(\tau) = (1-e^{-\kappa_x \tau})/\kappa_x. \qquad (30)$$

The second derivative of the forward rate curve with respect to B_x is

$$\frac{\partial^2 f}{\partial B_x^2} = \alpha(S)\frac{\partial^2 B_S}{\partial B_x^2} - \sigma_s^2\left[B_s\frac{\partial^2 B_s}{\partial B_x^2} + \left(\frac{\partial B_s}{\partial B_x}\right)^2\right] + \alpha(l)\frac{\partial^2 B_l}{\partial B_x^2} - \sigma_l^2\left[B_l\frac{\partial^2 B_l}{\partial B_x^2} + \left(\frac{\partial B_l}{\partial B_x}\right)^2\right]. \qquad (31)$$

Equation (31) depends on the derivatives of B_s and B_l with respect to B_x. For B_s these are

$$\partial B_s/\partial B_x = e^{-(\kappa_s-\kappa_x)\tau} \quad \text{and} \quad \partial^2 B_s/\partial B_x^2 = -(\kappa_s-\kappa_x)\,e^{-(\kappa_s-2\kappa_x)\tau}. \qquad (32)$$

The derivatives with respect to B_l are identical except that κ_l replaces κ_s. In general these functions may be either increasing or decreasing with maturity. However, consider the empirically relevant case where κ_l is close to zero, where κ_s is positive and in the range (say) 0.3–0.7 and where we set κ_x equal to κ_l. In this case, as maturity increases, the first and second derivatives of B_s tend towards zero, the second derivative of B_l (with respect to itself) is precisely zero while the first derivative is unity. As maturity increases, therefore, minus the curvature of the forward rate curve measured with respect to B_l tends towards the variance of the long rate. In other words we have

Proposition 6. *In the two factor model described in* (24), (25) *and* (26) *and for* $\kappa_s-\kappa_l > 0$ *and* $\kappa_s-2\kappa_l > 0$, *minus the second derivative of the forward rate with respect to* B_l *tends to the variance of the long rate,* σ_l^2, *as maturity tends to infinity.*

Figure 2 illustrates this result for (*a*) different values of κ_s and (*b*) different levels of the spread. In figure 2(*a*), κ_l is equal to its mean value in table 1 (0.0026) and the values of κ_s are typical of those reported in table 1. Here, for maturities longer than 10–15 years, minus the convexity of the forward rate curve converges to close to the variance of the long rate. The large divergences for shorter maturities are a result of the first term in (31), which is:

$$\alpha(S)\,\partial^2 B_s/\partial B_l^2.$$

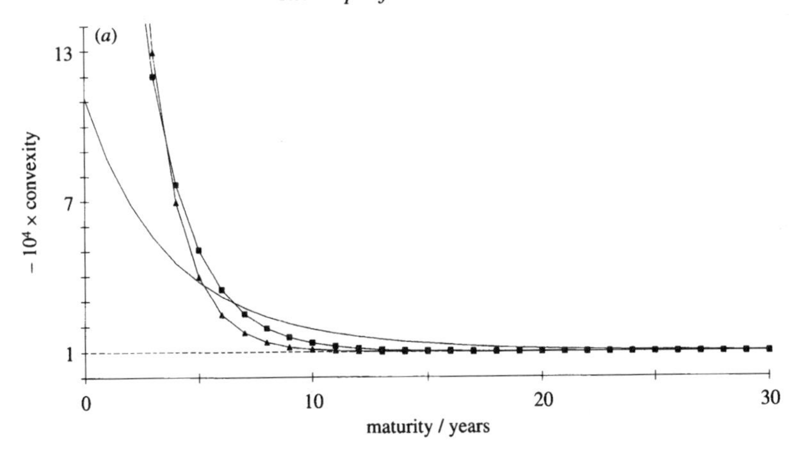

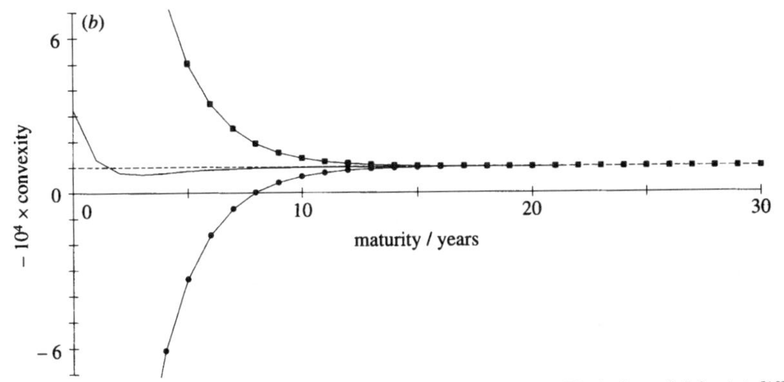

Figure 2. The convexity of the forward rate curve in the bivariate Vasicek model for (*a*) different levels of mean reversion in the spread, κ_S. Spread mean reversion coefficient values: ——, 0.2; —■—, 0.5; —▲—, 0.7; ······, variance of long rate ($\times 10^4$). Other parameter values are $\kappa_l = 0.0026$, $\mu_l = 0.05$, $\mu_S = 0.00$, s.d. $(l) = 100$ b.p., s.d. $(S) = 150$ b.p., $l = 0.05$ and $S = -0.02$. (*b*) Shows convexity for different values of the spread, S. —■—, spread $= -2\%$; ——, spread $= 0\%$; —●—, spread $= +2\%$; ······, variance of long rate ($\times 10^4$). Other parameters as (*a*) with $\kappa_S = 0.50$.

When S is close to its mean, and the first term in (31) is correspondingly small, figure 2(*b*) shows that the convexity of the forward rate is close to the variance of the long rate for almost all maturities.

(*b*) *A puzzle from the empirical literature*

Brown & Dybvig (1986) and Barone *et al.* (1991) fit the single factor CIR model to cross-sectional data on nominal bond prices. They find that in some important respects the model clearly fails; in particular, the model predicts that long term interest rates are constant whereas both studies report parameter estimates which are quite inconsistent with a constant long rate. None the less both studies appear to obtain quite 'reasonable' estimates of implied interest rate volatility. This is surprising because in both studies parameters are estimated from cross-sectional data only and the variability of long term interest rates should be reflected in the shape of the term structure. Yet, despite the fact that the model ignores uncertainty in the long rate, the implied interest rate volatility is of the correct order of magnitude.

Figure 2 may contain at least a partial explanation for this puzzle. As we have shown, estimating the variance in single factor affine yield models is equivalent to estimating the convexity of the forward rate curve with respect to the corresponding $B(\tau)$. Further, to the extent that the studies cited above estimate parameters from bonds of all maturities, we may regard the convexity estimate as a weighted average convexity across forward rates of different maturities. From figure 2 we see that, for maturities greater than 10 years, the convexity is almost constant and equal to the variance of the long rate. The variance estimates obtained by Brown & Dybvig (1986) and by Barone *et al.* (1991) may therefore reflect the variance of the long rate for longer maturities and, for shorter maturities, a certain amount of 'noise' associated with the sign of the spread.

7. Conclusions

We have analysed the influence of interest rate uncertainty on the shape of the yield curve. For a broad class of models – the time homogeneous affine yield class – we find that the forward rate is a relatively simple function of the parameters of the risk adjusted process for the short rate and a measure of the term structure of volatility or 'duration'. Here, for any maturity, minus the convexity of the forward rate curve with respect to the duration measure is equal to the variance of the short rate. In the time homogeneous single factor case, it was shown that a model belongs to the affine yield class if and only if the drift and diffusion coefficients of the risk adjusted process for the short rate are affine in r.

In the multivariate case, where Duffie & Kan (1993) have shown that an analogous result holds, each state variable is associated with a duration measure. However, empirical evidence suggests that (*a*) two factors account for the great majority of yield curve movements and (*b*) the maturity profile of the two associated duration measures differ significantly. In this case it was shown that minus the convexity of the forward rate curve with respect to the duration measure associated with the long rate converges to the variance of the long rate.

This research has been funded by the Institute for Quantitative Investment Research. Part of the work was carried out during the period the second named author spent at the Dipartimento di Scienze Economiche of the Università degli Studi di Venezia as the Cassa di Risparmio di Venezia Visiting Fellow. S.M.S. thanks his colleagues in the department for their warm hospitality and the bank for their generosity in funding the Fellowship. We are indebted to Farshid Jamshidian who pointed out an error in an earlier draft of this paper. The comments and suggestions of Darrell Duffie are gratefully acknowledged.

References

Ayres, H. R. & Barry, J. Y. 1979 The equilibrium yield curve for government securities. *Financial Analysts. J.* May/June, 31–39.

Ayres, H. R. & Barry, J. Y. 1980 A theory of the U.S. Treasury market equilibrium. *Management Sci.* **26**, 539–569.

Barone, E., Cuoco, D. & Zautzik, E. 1991 Term structure estimation using the Cox, Ingersoll and Ross model: the case of Italian Treasury Bonds. *J. Fixed Income* **1**, 87–95.

Brown, S. J. & Dybvig, P. H. 1986 The empirical implications of the Cox, Ingersoll, Ross theory of the term structure of interest rates. *J. Finance* **XLI**, 617–630.

Brown, R. H. & Schaefer, S. M. 1988 Testing the Cox, Ingersoll & Ross model on British government index-linked securities. Institute of Finance and Accounting, London Business School. (Working paper.)

Brown, R. H. & Schaefer, S. M. 1994 The term structure of real interest rates and the Cox, Ingersoll & Ross model. *J. Financial Econ.* **35**, 3–42.

Caverhill, A. & Strickland, C. 1992 Money market term structure dynamics and volatility expectations. Financial Options Research Centre, University of Warwick. (Working paper.)

Cox, J. C., Ingersoll, J. E. & Ross, S. A. 1979 Duration and the measurement of basis risk. *J. Business* **52**, 51–61.

Cox, J. C., Ingersoll, J. E. & Ross, S. A. 1980 A theory of the term structure of interest rates. Stanford University. (Working paper.)

Cox, J. C., Ingersoll, J. E. & Ross, S. A. 1981 A re-examination of tradition hypotheses about the term structure of interest rates. *J. Finance* **XXXVI**, 769–799.

Cox, J. C., Ingersoll, J. E. & Ross, S. A. 1985 A theory of the term structure of interest rates. *Econometrica* **53**, 385–407.

Duffie, D. & Kan, R. 1993 A yield-factor model of interest rates. Stanford University. (Working paper.)

Litterman, R. & Scheinkman, J. 1991 Common factors affecting bond returns. *J. Fixed Income* **1**, 54–61.

Litterman, R., Scheinkman, J. & Weiss, L. 1991 Volatility and the yield curve. *J. Fixed Income* **1**, 49–53.

Longstaff, F. A. & Schwartz, E. S. 1990 Interest rate volatility and the term structure: a two-factor general equilibrium model. *J. Finance* **47**, 1259–1282.

Nelson, C. & Schaefer, S. M. 1983 The dynamics of the term structure and alternative immunization strategies. In *Innovations in bond portfolio management: duration analysis and immunization* (ed. G. O. Bierwag, G. G. Kaufman and A. Toevs). Greenwich, CT: JAI Press.

Pearson, N. & Sun, T. 1988 A test of the CIR model of the term structure of interest rates using the method of maximum likelihood. Sloan School of Management, MIT. (Working paper.)

Merton, R. C. 1971 Optimum consumption and portfolio rules in a continuous-time model. *J. Econ. Theory* **3**, 373–413.

Schaefer, S. M. 1981 Measuring a tax-specific term structure of interest rates in the market for British government securities. *Econ. J.* **91**, 415–423.

Schaefer, S. M. & Schwartz, E. S. 1984 A two-factor model of the term structure: an approximate analytical solution. *J. Financial Quant. Anal.* **19**, 413–424.

Vasicek, O. 1977 An equilibrium characterization of the term structure. *J. Financial Econ.* **5**, 177–188.

Discussion

L. C. G. ROGERS (*Queen Mary & Westfield College, London University, U.K.*). In a working paper, 'Long forward and zero-coupon rates can never fall', Dybvig, Ingersoll and Ross (DIR) prove that if the long rate process l_t exists, then necessarily it is an increasing process. (The long rate is defined by $l_t = \lim_{T-\infty} -T^{-1} \log P(t, T)$, where $P(t, T)$ is the price at time t of a zero-coupon bond which delivers 1 at a later time T.) Thus any model in which the long rate is allowed to decrease must offer the possibility of arbitrage, and this appears to be a fundamental objection to the model of Schaefer and Schwartz.

S. M. SCHAEFER. The model proposed in my paper with Schwartz uses the yield on a *consol*, a bond with infinite maturity paying a regular coupon, rather than the infinite maturity *zero-coupon yield*, as the second state variable and stochastic variation in the former is not inconsistent with DIR's position, e.g. the infinite maturity zero-coupon yield being a constant.

S. MEHTA (*London, U.K.*). One model of the term structure is to regard the yield on a long-term bond as the sum of expected future short term interest rates plus a term premium. I would be interested in your comments on the level of term premium.

S. M. SCHAEFER. Empirical evidence on the level of the term premium is weak: typical premia in rates of return on long term government bonds are around 1%, but the standard error of this estimate is of the same order. Modern 'factor' models of the term structure, including those described in our paper, provide useful cross-sectional restrictions on term premia but, unfortunately, little guidance as to their level.

J. PEZIER (*Mitsubishi Finance International plc, London, U.K.*). Professor Schaefer has shown that when conducting a principal component analysis of the possible deformations of interest rate term structures in US dollars and £ sterling, 90% of the variance may well be explained by the first component and about 5% by the second component, leaving very little to be explained by higher order components. We certainly can corroborate these findings on the dozen major currencies we have analysed over the last 10 years. He went on to say that, similarly, one would need perhaps '2.1 parameters' in an equilibrium model of interest rate term structure, to achieve a realistic description of term structure evolution, provided the parameters are well chosen.

What, then, would be Professor Schaefer's recommendation to regulators who base risk measurements on the 'bucketing' of cash flows into as many as 20 different maturity bands and on adding the absolute interest rate exposures found in these bounds without any interband netting?

Practitioners recognize the importance of measuring risk and hedging positions in a robust manner, that is, in a manner that is not too dependent on a particular choice of model. Would not Professor Schaefer say, however, that 20 interest rate maturities are too many parameters and not the right ones?

S. M. SCHAEFER. I find little to disagree with in what Dr Pezier says. In particular, I agree that it is unlikely that capital adequacy rules for fixed income instruments actually need to monitor exposure to as many as 20 maturity bands. More generally, as I suggested in Schaefer (1992), capital adequacy rules would be much improved if they were much more firmly based on empirical analysis.

Additional reference

Schaefer, S. M. 1992 Financial regulation: the contribution of the theory of finance. In *The internationalisation of capital markets and regulatory response* (ed. J. Fingleton). Graham & Trotham.

12

Multi-factor term structure models

DARRELL DUFFIE AND RUI KAN

This is a survey of multi-factor structure models, concentrating on models in which the term structure has a finite-dimensional (Markov diffusion) state-space representation. The special 'affine' case is shown to be tractable.

1. Introduction

Figure 1 shows the term structure of interest rates at a given point in time. Stochastic models for fluctuations in term structure over time are commonly used in the finance industry for at least the following purposes.

(i) The pricing of fixed-income derivative securities, such as options and mortgage-backed securities.

(ii) The analysis of the risk of fixed-income portfolio strategies.

(iii) Managing the interest-rate risk of fixed-income positions.

By 'fixed income', we mean assets whose pay-offs depend on the term structure itself. In a wide sense, this can include bonds, bond derivatives such as options, swaps, or caps, defaultable bonds, and even foreign bonds or derivatives based in sometimes complicated ways on domestic and foreign interest rates. There are many other reasons for understanding the process by which interest rates are determined and change over time, but our focus will be on models that are particularly useful for the above three purposes.

Although various classes of stochastic models are used, the most common language of term structure modellers in industry and universities is that of continuous-time stochastic calculus, which reached popularity following the impact of the Black & Scholes (1973) option pricing formula and the associated modelling ideas developed by Merton (1973) and others. We will review how such models are constructed and applied, with particular reference to Markov diffusions that represent the current term structure in a finite-dimensional state space. Within this class, one can make reasonable trade-offs between economic realism and computational tractability, bearing in mind that no tractable model can fully capture the complexity of unexpected changes in interest rates.

2. Setup

We begin with a probability space $(\Omega, \mathcal{F}, P)$ and the augmented filtration $\{\mathcal{F}_t : t \in [0, \infty)\}$ generated by a standard brownian motion W^* in $\mathbb{R}^n$, for some $n \geqslant 1$. (For technical details, see, for example, Karatzas & Shreve (1988), Protter (1990) or other standard references.)

Given is a progressively measurable 'short rate' process r such that $\int_0^T |r_t| \, \mathrm{d}t < \infty$ almost surely for all $T > 0$. We may think of r_t as the interest rate at time t on loans

Mathematical Models in Finance. Edited by S.D. Howison, F.P. Kelly and P. Wilmott.
Published in 1995 by Chapman & Hall, London. ISBN 0 412 63070 2

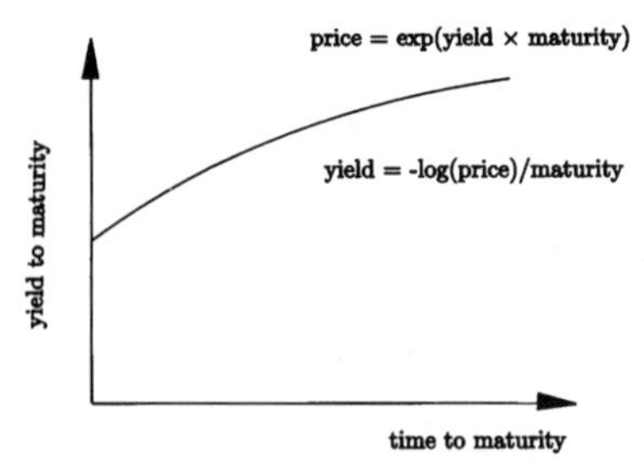

Figure 1. The yield curve.

of infinitesimal maturity. More properly, it is possible to invest one unit of account at any time t in deposits and receive at any $s \geqslant t$ the pay-off $\exp(\int_t^s r_u \, du)$.

For purposes of this survey, a security is a financial claim promising, for some time T, a pay-off defined by some $\mathscr{F}_T$-measurable random variable u. According to a model of Harrison & Kreps (1979), as subsequently developed by many (see, for example, the references cited in Duffie 1992, especially Ansel & Stricker 1992), under technical conditions there is not arbitrage if and only if there is a probability measure Q, equivalent to P, under which the price of a security paying u at time $T \geqslant t$ is given by

$$E\left[\exp\left(-\int_t^T r_s \, ds\right) u \,\middle|\, \mathscr{F}_t\right]. \tag{1}$$

Here, and throughout, E denotes expectation under such a probability measure Q, which is fixed. The obvious example is to take $u = 1$, defining the price $p_{t,T}$ of a zero-coupon bond maturing at T. The continuously compounding yield of a bond of maturity τ is then defined as

$$y_{t,\tau} = (-1/\tau) \log p_{t,t+\tau}, \quad (t, \tau) \in \mathbb{R}_+^2. \tag{2}$$

For practical applications, there remains the basic issue of how to model the probabilistic behaviour of the short rate process r under Q. One wants a model for r that is sufficiently rich to capture the essential nature of the actual market, while at the same time sufficiently tractable for purposes of econometric estimation and for computation of the prices of contingent claims as in (1), for a range of commonly traded securities whose pay-offs are represented by u. There are also many theoretically interesting questions regarding the equilibrium determination of the short rate process r and the equivalent 'martingale' measure Q. It is known that, under weak technical conditions, any short rate process r can be supported in a simple general equilibrium setting with easily specified utility functions and consumption endowments (see, for example, Heston 1991; Duffie 1992, exercise 9.3). In any case, we will be focusing here only on practical issues, and disregarding other aspects of the general equilibrium problem. From this point, we will review some basic classes of models for the behaviour of the short rate process r under the equivalent martingale measure Q. We begin with 'single-factor' models, move to 'multi-factor' models, and finally describe 'infinite-factor' models in the framework of Heath *et al.* (1992). For many applications, it will also be useful to model the distribution of processes under the original probability measure P. Conversion from P to Q and back will not be dealt with here, but is an important issue, particularly

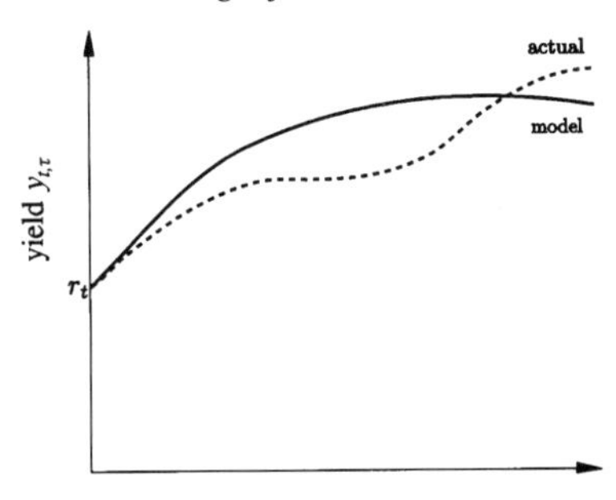

Figure 2. Actual and modelled yield curves.

from the point of view of statistical fitting of the models as well as the measurement of risk.

3. Single-factor models

The simplest class of models that we consider takes the short rate process to be the solution of a stochastic differential equation of the form

$$dr_t = \mu(r_t)\,dt + \sigma(r_t)\,dW_t, \tag{3}$$

where W is a standard brownian motion under Q and where $\mu: \mathbb{R} \to \mathbb{R}$ and $\sigma: \mathbb{R} \to \mathbb{R}$ have enough regularity to ensure the existence of a unique solution to (3) (see, for example, Ikeda & Watanabe 1981). Since r is a strong Markov process under Q, we have $p_{t,T} = F(r_t, t)$, for some measurable function $F: \mathbb{R} \times [0, T] \to \mathbb{R}$, and we can therefore view the entire yield curve $y_t = \{y_{t,\tau}: \tau \geqslant 0\}$ defined by (2) as measurable with respect to r_t. Hence the label 'single-factor model' applies, since a single state variable, in this case the short rate r_t, is a sufficient statistic for all future yield curves.

Although simple and, as it turns out, quite tractable, the single-factor class of models given by (3) is (like any theoretical model) at variance with reality. Consequently, on a given day, the yield curve associated with the model differs from that observed in the market-place, as depicted in figure 2.

If significant, this discrepancy may suggest the development of a new theoretical model. In the finance industry, however, one needs to use some particular model, even if it is imperfect. In practice, the discrepancy between the actual and theoretical yield curves depicted in figure 2 is eliminated by introducing at the current time t, time dependence in the functions μ and σ, to arrive at a 'calibrated model' $\mu^t: \mathbb{R} \times [0, \infty) \to \mathbb{R}$ and $\sigma^t: \mathbb{R} \times [0, \infty) \to \mathbb{R}$, of the form

$$dr_s = \mu^t(r_s, s)\,ds + \sigma^t(r_s, s)\,dW_s, \quad s \geqslant t. \tag{4}$$

This calibrated model (μ^t, σ^t) is computed numerically from the original model (μ, σ) using algorithms that are described, for example, in Black *et al.* (1990). With proper calibration, the result is an exact match between the actual and modelled yield curves. Indeed, it is common to calibrate not only with the current yield curve, but also with certain volatility-related information available in the market through the prices of options.

At the next time period $t+1$, of course, there is again a discrepancy between the observed market yield curve and the yield curve computed at the new short rate r_{t+1}

Table 1. *Parametric single-factor models*

$$dr_t = [\alpha_1(t) + \alpha_2(t)\,r_t + \alpha_3(t)\,r_t \log r_t]\,dt + [\beta_1(t) + \beta_2(t)\,r_t]^\gamma\,dB_t$$

	α_1	α_2	α_3	β_1	β_2	γ
Cox–Ingersoll–Ross	•	•	—	—	•	0.5
Dothan	—	—	—	—	•	1.0
Brennan–Schwartz	•	•	—	—	•	1.0
Merton	•	—	—	•	—	1.0
Vasicek	•	•	—	•	—	1.0
Pearson–Sun	•	•	—	•	•	0.5
Black–Derman–Toy	—	•	•	—	•	1.0
Constantinides–Ingersoll	—	—	—	—	•	1.5

with the previous calibration (μ^t, σ^t) of the model. It is common in practice to re-calibrate to a new model $(\mu^{t+1}, \sigma^{t+1})$. Since the necessity for re-calibration was not considered when using the previous version of the model for pricing purposes, this suggests a theoretical inconsistency in the application of the model. The compromise involved seems reasonable under the circumstances. It has sometimes been said that one can avoid this compromise with the modelling approach of Heath *et al.* (1992), since that framework essentially admits an arbitrary initial yield curve without the need for calibration. In effect, the state variable for the Heath–Jarrow–Morton (HJM) model is the entire yield curve itself. In fact, the HJM model admits movements in the yield curve generated only by a finite-dimensional brownian motion and therefore limits the sorts of movements of the yield curve that can be considered without calibration. Recent work by Kennedy (1992), however, extends the HJM model to allow for an infinite-dimensional brownian motion (in the framework of stochastic flows).

Most, if not all, of the parametric single-factor models appearing in the literature or in industry practice, are of the form

$$dr_t = [\alpha_1(t) + \alpha_2(t)\,r_t + \alpha_3(t)\,r_t \log r_t]\,dt + [\beta_1(t) + \beta_2(t)\,r_t]^\gamma\,dW_t, \tag{5}$$

for time-dependent deterministic coefficients α_1, α_2, α_3, β_1, and β_2, and for some exponent $\gamma \geqslant 0.5$. (For existence and uniqueness of solutions, additional coefficient restrictions apply.) Table 1 lists the origins of various special cases of this parametric class, indicating with • the coefficients that are non-zero (sometimes constant) for each special case, and indicating the choice of power γ. (By offering extensions with time-varying coefficients, Ho & Lee (1986) and Hull & White (1990) have popularized the constant coefficients models of Merton (1973) and Vasicek (1977).)

Even for this simple parametric class (5), there are clearly degrees of freedom in calibrating the model to the observed yield curve. It is also common in practice to calibrate the model to market prices for derivative securities, such as bond options or 'caps', both of which provide useful volatility-related information that can be used to obtain more realistic model behaviour. For a description, see Black & Karasinski (1992), for example.

4. Multi-factor models

Although single-factor models offer tractability, there is compelling reason to believe that a single state variable, such as the short rate r_t, is insufficient to capture reasonably well the distribution of future yield curve changes. The econometric evidence in favour of this view includes the work of Litterman & Scheinkman (1988), Stambaugh (1988), Pearson & Sun (1990), and Chen & Scott (1992*b*, 1993). (For empirical comparisons among most of the single-factor models considered in table 1, see Chan *et al.* (1992).)

In principle, of course, the yield curve sits in an infinite-dimensional space of functions, and there is no reason to believe that the direction of its movements will be restricted to some finite-dimensional manifold. For practical purposes, however, tractability suggests a finite number of state variables, and it is an empirical issue as to how many might be sufficient to offer reasonable empirical properties. Some of the empirical studies mentioned above suggest that 2 or 3 state variables might suffice for practical purposes.

In any case, we will consider a state process X in some open subset D of $\mathbb{R}^n$, defined as the solution to

$$\mathrm{d}X_t = \mu(X_t)\,\mathrm{d}t + \sigma(X_t)\,\mathrm{d}W_t, \tag{6}$$

where W is a standard brownian motion in $\mathbb{R}^n$ under Q, and where $\mu: D \to \mathbb{R}^n$ and $\sigma: D \to \mathbb{R}^{n \times n}$ satisfy sufficient regularity for existence and uniqueness of solutions. In what follows, we could add time dependence to μ and σ without changing the major ideas.

We also suppose that the short rate process r is given by $r_t = R(X_t)$, for some $R: D^n \to \mathbb{R}$. Thus the zero-coupon bond maturing at T has a price at $t \leqslant T$ given from (1) by

$$F(X_t, t) = E\left[\exp\left(-\int_t^T R(X_s)\,\mathrm{d}s\right)\middle| X_t\right]. \tag{7}$$

One could imagine that the state vector X_t might include various economic indices that would affect interest rates such as economic activity, monetary supply variables, central bank policy objectives, and so on. In order to facilitate the pricing and hedging of fixed-income derivatives, however, it is convenient to assume that one can find a change of variables under which we may view X_t as yield-related variables. This will be one of our objectives. We also desire a model that has some measure of numerical and econometric tractability. For both of these reasons, it may turn out to be convenient to take μ, $\sigma\sigma^{\mathrm{T}}$ and R to be affine functions on D into their respective ranges. (An affine function is a constant plus a linear function.) In this case, we say that the primitive model $(\mu, \sigma\sigma^{\mathrm{T}}, R)$ is affine.

Likewise, we say that the term structure is itself affine if there are C^1 functions $c: [0, \infty) \to \mathbb{R}$ and $C: [0, \infty) \to \mathbb{R}^n$ such that

$$y_{t,\tau} = c(\tau) + C(\tau)\cdot X_t, \quad t \geqslant 0, \quad \tau \geqslant 0, \tag{8}$$

so that yields are affine in the state variables.

Indeed, in Duffie & Kan (1992) it is shown that, under technical conditions, the basic model $(\mu, \sigma\sigma^{\mathrm{T}}, R)$ is affine if and only if the term structure is affine. This extends the same result for $n = 1$ given by Brown & Schaefer (1991). For an affine model, Duffie & Kan (1992) show that the coefficient functions c and C of (8) solve an ordinary differential equation of the form

$$C_i'(\tau) = k_i + K_i \cdot C(\tau) + C(\tau)^{\mathrm{T}} Q_i\, C(\tau), \quad i \in \{1, \ldots, n\}, \tag{9}$$

$$c'(\tau) = k_0 + K_0 \cdot C(\tau) + C(\tau)^{\mathrm{T}} Q_0\, C(\tau), \tag{10}$$

with boundary conditions

$$c(0) = C_i(0) = 0, \quad i \in \{1, \ldots, n\}, \tag{11}$$

where $\{k_0, \ldots, k_n\} \subset \mathbb{R}$, $\{K_0, \ldots, K_n\} \subset \mathbb{R}^n$, and $\{Q_0, \ldots, Q_n\} \subset \mathbb{R}^{n\times n}$ are constant coefficients given in terms of the coefficients defining the underlying affine functions μ, $\sigma\sigma^{\mathrm{T}}$, and R. The Ricatti equation (9)–(11) can easily be solved numerically, for example by a Runge–Kutta method.

Given the solution (c, C) of (9)–(11), relation (8) provides an affine change of variables under which the state may be taken to be an n-dimensional 'yield-factor' process Y, where for some fixed maturities $\tau(1), \ldots, \tau(n)$, we take

$$Y_{ti} = y_{t,\tau(i)} = c(\tau(i)) + C(\tau(i)) \cdot X_t, \quad i \in \{1, \ldots, n\}. \tag{12}$$

We need only ensure that the 'basis maturities' $\tau(1), \ldots, \tau(n)$ are chosen so that the matrix K in $\mathbb{R}^{n\times n}$, defined by $K_{ij} = C_j(\tau(i))$, is non-singular. In that case, we have $Y_t = k + KX_t$, where $k_i = c(\tau(i))$, and the new state dynamics are given by

$$\mathrm{d}Y_t = \mu^*(Y_t)\,\mathrm{d}t + \sigma^*(Y_t)\,\mathrm{d}Wt, \tag{13}$$

where

$$\mu^*(y) = k\mu(K^{-1}y - k),$$

$$\sigma^*(y) = K\sigma(K^{-1}y - k),$$

for $y \in D^* = \{Kx + k \colon x \in D\}$.

If σ is constant, X and Y are Gauss–Markov processes of the Ornstein–Uhlenbeck form. For abstract factors, this gaussian model was developed by Langetieg (1980) and Jamshidian (1990, 1991). A Gauss–Markov yield-factor model was developed by El Karoui & Lacoste (1992) in the forward-rate setting of Heath *et al.* (1992), and in the current state-space setting, was developed as a special case of stochastic volatility models by Duffie & Kan (1992).

A simple example of non-constant σ is the multivariate Cox–Ingersoll–Ross model:

$$\mathrm{d}X_{it} = (a_i - b_i X_{it})\,\mathrm{d}t + c_i \sqrt{X_{it}}\,\mathrm{d}W_{it}, \quad i \in \{1, \ldots, n\}, \tag{14}$$

for positive constants a_i, b_i, c_i, appearing in Feller (1951), and developed for interest-rate modelling by Cox *et al.* (1985), Richard (1978), Heston (1991), Longstaff & Schwartz (1992), and Chen & Scott (1992*a*). Restrictions apply. For all i, we want

$$a_i > c_i^2/2. \tag{15}$$

As shown by Ikeda & Watanabe (1981), the latter restriction is necessary and sufficient to ensure that X will remain in the obvious open state space $D = \mathrm{int}(\mathbb{R}^n_+)$. Duffie & Kan (1992) study the general case, under which we can without loss of generality take

$$\mu(x) = ax + b; \quad \sigma_{ij}(x) = \gamma_{ij} \sqrt{(\alpha_{ij} + \beta_{ij} \cdot x)}, \tag{16}$$

for some $\gamma_{ij} \in \mathbb{R}$, $\alpha_{ij} \in \mathbb{R}$, $\beta_{ij} \in \mathbb{R}^n$, $a \in \mathbb{R}^{n\times n}$, and $b \in \mathbb{R}^n$. In this case, the state space is

$$D = \{x \in \mathbb{R}^n \colon \alpha_{ij} + \beta_{ij} \cdot x > 0, \quad i, j \in \{1, \ldots, n\}\}. \tag{17}$$

Strong restrictions on the coefficients $(a, b, \gamma, \alpha, \beta)$, analogous to (15) but more complicated, are shown by Duffie & Kan (1992) to imply the affine form and to guarantee the existence and uniqueness of solutions to $\mathrm{d}X_t = \mu(X_t)\,\mathrm{d}t + \sigma(X_t)\,\mathrm{d}W_t, x_0 \in D$, for (16)–(17).

Aside from the affine case, multivariate term-structure models appear in Brennan & Schwartz (1979), Chan (1992), El Karoui *et al.* (1992), Constantinides (1992), Beaglehole & Tenney (1991) and Jamshidian (1993). Most of these non-affine multifactor models do not allow direct observation of the state from the yield curve.

If one does not observe the state-vector directly, in principle one can filter the state variable from yield-curve data. There are debates concerning how much this limited observation property detracts from the practical application of the models. It can be said, for example, that we do not observe the yield curve in any case, but merely the prices of coupon bonds, from which one infers statistically (and with noise) the zero-coupon curve by some curve-fitting method such as splines or nonlinear least squares. In any case, it seems to be of at least some value to have state variables that can be observed in terms of the yield curve, as in the affine models described above.

5. Derivative pricing

Given a term structure model (μ, $\sigma\sigma^{\mathrm{T}}$, R), affine or not, one is interested in the pricing of derivative securities. Recall that the price of a security with pay-off u at time T is given at time t by

$$E\left[\exp\left(-\int_t^T R(X_s)\,\mathrm{d}s\right)u\,\bigg|\,\mathscr{F}_t\right].$$

If u is measurable with respect to the yield curve at time T, as are bond options and other 'path-independent' derivatives, we may take $u = g(X_T)$ for some $g: D^n \to \mathbb{R}$, since the yield curve y_T is itself X_T-measurable. In this case, the Markov property of X implies that we can write the derivative price as

$$F(X_t, t) = E\left[\exp\left(-\int_t^T R(X_s)\,\mathrm{d}s\right)g(X_t)\,\bigg|\,X_t\right], \tag{18}$$

for some $F: D \times [0, T] \to \mathbb{R}$. Under the technical regularity given, for example, in Friedman (1975), we also know that F is the unique solution in $C^{2,1}(D^n \times [0, T))$, under technical growth conditions, to the parabolic partial differential equation

$$\mathscr{D}F(x,t) - R(x)\,F(x,t) = 0, \quad (x,t) \in D \times [0, T), \tag{19}$$

$$F(x,t) = g(x), \quad x \in D, \tag{20}$$

where

$$\mathscr{D}f(x,t) = F_x(x,t)\,\mu(x) + F_t(x,t) + \tfrac{1}{2}\,\mathrm{trace}\,(\sigma(x)\,\sigma(x)^{\mathrm{T}}\,F_{xx}(x,t)).$$

One can then solve for path-independent derivative prices via a numerical solution of the partial differential equation (19)–(20), say by finite-difference methods. (For finite-difference algorithms, see for example Ames (1977).) For affine multi-factor models, fully worked examples are given by Duffie & Kan (1992) for the case $n = 2$. For large n, say more than 3, currently available algorithms and hardware are not up to the task, and Monte Carlo simulation may be applied (see, for example, Duffie & Glynn 1992; Kloeden & Platen 1992). For the path-dependent case, unless there is a simple way to augment the state space so as to capture the path dependence with an additional state variable, it may also be advisable to resort to Monte Carlo simulation. There are only rare cases, such as Jamshidian's (1991) solution for bond options in the gaussian setting, for which one can obtain explicit solution for derivative prices (see, also, El Karoui & Rochet 1989).

6. Where do we go from here?

A great deal of work remains to be done. First, we have discussed only the case of single-currency yield curves with no default risk. International models, which consistently include random exchange rate fluctuations, are difficult to model in a tractable way. An example, in the same affine state space setting emphasized here, is offered by Nielson & Saá-Requejo (1992). Modelling default risk in a consistent way, while maintaining tractability, is also challenging. Madan & Unal (1992) and Jarrow & Turnbull (1992) offer examples of models that push in this direction.

Econometric modelling of the term structure, particularly in a multi-factor setting, has stayed within a relatively narrow framework. Recent work by Gibbons & Ramaswamy (1992), Pearson & Sun (1990), and Chen & Scott (1992*b*, 1993), for example, stays strictly within the CIR single-factor or multi-factor cases of the affine model emphasized here. For the constant-volatility Gauss–Markov (affine) case, Frachot *et al.* (1992) together with Frachot & Lesne (1993) have done some empirical work in the Heath–Jarrow–Morton setting. Much remains to be done in integrating the use of statistical models within the practical applications of term structure models mentioned in the introduction.

Judging from the literature on term structure modelling, much also remains to be done in the development and application of numerical methods, such as finite-difference or finite-element algorithms for multidimensional Cauchy problems such as (19)–(20), to the particular sorts of applications that are found in fixed-income markets.

References

Ames, W. 1977 *Numerical methods for partial differential equations*, 2nd edn. New York: Academic Press.

Ansel, J.-P. & Stricker, C. 1991 Lois de Martingale, densités et décompositions de föllmer schweizer. Université de Franche-Comté. (Working paper.)

Beaglehole, D. & Tenney, M. 1991 General solutions of some interest rate contingent claim pricing equations. *J. Fixed Income* **1**, 69–83.

Black, F., Derman, E. & Toy, W. 1990 A one-factor model of interest rates and its application to treasury bond options. *Financial Analysts J.* 33–39.

Black, F. & Karasinski, P. 1992 Bonds and option pricing when short rates are log-normal. *Financial Analysts J.* 52–59.

Black, F. & Scholes, M. 1973 The pricing of options and corporate liabilities. *J. Political Econ.* **81**, 637–654.

Brennan, M. & Schwartz, E. 1979 A continuous time approach to the pricing of bonds. *J. Banking Finance* **3**, 133–155.

Brown, R. & Schaefer, S. 1991 Interest rate volatility and the term structure of interest rates. London Business School. (Working paper.)

Chan, K.-C., Karolyi, G., Longstaff, F. & Sanders, A. 1992 An empirical comparison of alternative models of the short-term interest rate. *J. Finance* **47**, 1209–1227.

Chan, Y.-K. 1992 *Term structure as a second order dynamical system and pricing of derivative securities.* Bear Stearns and Company.

Chen, R.-R. & Scott, L. 1992*a* Pricing interest rate options in a two-factor Cox–Ingersoll–Ross model of the term structure. *Rev. Financial Stud.* **5**, 613–636.

Chen, R.-R. & Scott, L. 1992*b* Maximum likelihood estimation for a multi-factor equilibrium model of the term structure of interest rates. Rutgers University and University of Georgia. (Working paper).

Chen, R.-R. & Scott, L. 1993 Multi-factor Cox–Ingersoll–Ross models of the term structure: estimates and tests from a state-space model using a Kalman filter. Rutgers University and University of Georgia. (Working paper.)

Constantinides, G. 1992 A theory of the nominal structure of interest rates. *Rev. Financial Stud.* **5**, 531–552.

Constantinides, G. & Ingersoll, J. 1984 Optimal bond trading with personal taxes. *J. Financial Econ.* **13**, 299–335.

Cox, J., Ingersoll, J. & Ross, S. 1985 A theory of the term structure of interest rates. *Econometrica* **53**, 385–408.

Dothan, M. 1978 On the term structure of interest rates. *J. Financial Econ.* **7**, 229–264.

Duffie, D. 1992 *Dynamic asset pricing theory.* Princeton University Press.

Duffie, D. & Glynn, P. 1991 Efficient Monte Carlo estimation of security prices. Graduate School of Business and Department of Operations Research, Stanford University. (Working paper.)

Duffie, D. & Kan, R. 1992 A yield-factor model of interest rates. Graduate School of Business, Stanford University. (Working paper.)

El Karoui, N. & Lacoste, V. 1992 Multi-factor models of the term structure of interest rates. University of Paris VI. (Working paper.)

El Karoui, N., Myneni, R. & Viswanathan, R. 1992 Arbitrage pricing and hedging of interest rate claims with state variables: I Theory. University of Paris VI. (Working paper.)

El Karoui, N. & Rochet, J.-C. 1989 A pricing formula for options on coupon bonds. University of Paris VI. (Working paper.)

Feller, W. F. 1951 Two singular diffusion problems. *Ann. of Math.* **54**, 173–182.

Frachot, A., Janci, D. & Lacoste, V. 1992 Factor analysis of the term structure: a probabilistic approach. Banque de France. (Working paper.)

Frachot, A. & Lesne, J.-P. 1993 Econometrics of linear factor models of interest rates. Banque de France. (Working paper.)

Friedman, A. 1975 *Stochastic differential equations and applications, Vol.* 1. New York: Academic Press.

Gibbons, M. & Ramaswamy, K. 1992 A test of the Cox–Ingersoll–Ross model of the term structure. Wharton School, University of Pennsylvania. (Working paper.)

Harrison, M. & Kreps, D. 1979 Martingales and arbitrage in multi-period security markets. *J. Econ. Th.* **20**, 381–408.

Heath, D., Jarrow, R. & Morton, A. 1992 Bond pricing and the term structure of interest rates: a new methodology for contingent claims valuation. *Econometrica* **60**, 77–105.

Heston, S. 1991 Testing continuous-time models of the term structure of interest rates. School of Organization and Management, Yale University. (Working paper.)

Ho, T. & Lee, S. 1986 Term structure movements and pricing interest rate contingent claims. *J. Finance* **41**, 1011–1029.

Huang, C.-F. 1987 An intertemporal general equilibrium asset pricing model: the case of diffusion information. *Econometrica* **55**, 117–142.

Hull, J. & White, A. 1990 Pricing interest rate derivative securities. *Rev. Financial Stud.* **3**, 573–592.

Ikeda, N. & Watanabe, S. 1981 *Stochastic differential equations and diffusion processes.* Amsterdam: North-Holland.

Jamshidian, F. 1990 The preference-free determination of bond and option prices from the spot interest rate. *Advances Futures Options Res.* **4**, 51–67.

Jamshidian, F. 1991 Bond and option evaluation in the gaussian interest rate model and implementation. *Res. Finance* **9**, 131–170.

Jamshidian, F. 1993 Bond, futures and option evaluation in the quadratic interest rate model. Fuji Bank, London. (Working paper.)

Jarrow, R. & Turnbull, S. 1992 Interest rate risk management in the presence of default risk. Cornell University, Graduate School of Management. (Working paper.)

Karatzas, I. & Shreve, S. 1988 *Brownian motion and stochastic calculus.* Springer-Verlag.

Kennedy, D. P. 1992 The term structure of interest rates as a gaussian random field. Statistical Laboratory, University of Cambridge. (Working paper.)

Kloeden, P. & Platen, E. 1992 *Numerical solution of stochastic differential equations.* Springer-Verlag.

Langetieg, T. 1980 A multivariate model of the term structure of interest rates. *J. Fin.* **35**, 71–97.

Litterman, R. & Scheinkman, J. 1988 Common factors affecting bond returns. Goldman Sachs Financial Strategies Group. (Research paper.)

Longstaff, F. & Schwartz, E. 1992 Interest rate volatility and the term structure: a two-factor general equilibrium model. *J. Finance* **47**, 1259–1282.

Madan, D. & Unal, H. 1992 Pricing the risks of default. University of Maryland, Department of Finance. (Working paper.)

Merton, R. 1973 Theory of rational option pricing. *Bell J. Econ. Man. Sci.* **4**, 141–183.

Nielsen, L. T. & Saá-Requejo, J. 1992 Exchange rate and term structure dynamics and the pricing of derivative securities. INSEAD. (Working paper.)

Pearson, N. & Sun, T.-S. 1990 An empirical examination of the Cox–Ingersoll–Ross model of the term structure of interest rates using the method of maximum likelihood. Graduate School of Business, Columbia University. (Working paper.)

Pennachi, G. 1991 Identifying the dynamics of real interest rates and inflation: evidence using survey data. *Rev. Financial Stud.* **4**, 53–86.

Protter, P. 1990 *Stochastic integration and differential equations.* New York: Springer-Verlag.

Richard, S. 1978 An arbitrage model of the term structure of interest rates. *J. Financial Econ.* **6**, 33–57.

Stambaugh, R. 1988 The information in forward rates: implications for models of the term structure. *J. Financial Econ.* **21**, 41–70.

Vasicek, O. 1977 An equilibrium characterization of the term structure. *J. Financial Econ.* **5**, 177–188.

13

Dynamic asset allocation: insights from theory

STEWART D. HODGES

This paper provides a survey of the now considerable academic theory relating to the practice of dynamic asset allocation. This work is scattered through the literature and many of the key ideas are not as accessible or well known as they deserve to be.

The paper begins by providing a definition of what is meant by dynamic asset allocation and a description of its most significant features. Next it develops the concept of path independence and its relationship to efficient diversification through time. It is shown that this principle also applies to funds whose performance is appraised relative to an index benchmark. The final sections of the paper describe the implications of recent work on market equilibrium and on performance measurement.

---

1. Introduction

There is now a substantial body of academic theory relating to the issues involved in dynamic asset allocation. However, this work is rather scattered through the literature and not very easy to find. This paper provides a survey of key ideas in the area. A number of these are not as accessible or as well known as they deserve to be.

The paper begins by providing a definition of what we mean by dynamic asset allocation and some of the most significant features of it. Section 3 discusses the concept of diversification through time. This is a very much less familiar concept than that of diversification across securities. Many funds measure performance relative to benchmarks, so §4 discusses problems related to setting objectives in terms of index benchmarks. Section 5 considers the role of market regularities and equilibrium including the use of the role of forecasts and tactical asset allocation. Finally in §6 we look at issues related to performance measurement.

2. Horizon distributions and contingent pay-offs

It is convenient to start from the definition of dynamic asset allocation given by Trippi & Harrif (1991). They define dynamic asset allocation as 'a class of investment strategies that shifts the content of portfolios between two or more asset classes in response either to changes in the value of the portfolio and/or external economic states, on a more or less continual basis'. The motivation is two-fold: first, to tailor the distribution of fund return at some future date so that it can be an entirely different shape from that of the market index. It can be skewed to the right or skewed to the left or tailored in other more ingenious ways. Second, it may be to exploit predictable regularities, which include market timing and other tactical allocation strategies.

In 1973, Black & Scholes discovered that options on assets can be replicated, and hence valued by shifting the content of a portfolio dynamically between two or more

Mathematical Models in Finance. Edited by S.D. Howison, F.P. Kelly and P. Wilmott.
Published in 1995 by Chapman & Hall, London. ISBN 0 412 63070 2

asset classes. A large body of work on option theory has resulted and it is to this that we must turn in order to understand what dynamic asset allocation strategies are all about. Option theory tells us how to value and synthesize a given contingent profile. The Black–Scholes formula tells us the value of the simplest of contingent profiles, the pay-offs from a call option, for example on a market index. Black–Scholes theory tells us not only how much money we need if we want to construct one of those, but also how to construct it. Provided the assumptions of Black & Scholes hold, the delta hedging strategy will exactly synthesize the contingent pay-offs of the option.

If we have a more complicated contingent profile we can always think of it as being built up in terms of a series of call options. Suppose we can approximate the profile by a series of straight lines. Then corresponding to the intercept we need an investment in that amount of zero coupon bonds. We will then achieve this intercept value even if the index falls to zero. To get the correct slope at this point, we need a corresponding amount in stocks. Finally, every time we have a change in the slope, the amount by which the slope changes tells us how many extra call options we have to have at that particular strike price. Thus any piecewise linear profile is very easy to build up in terms of call options which are easy to value and we know how to replicate using the Black–Scholes theory.

For any fixed investment horizon, the profile of the fund value contingent on possible different values of the market index is exactly equivalent to the probability distribution of the fund value. We can always think in terms of mapping backwards or forwards between a probability distribution and an equivalent contingent profile. To do this we first need to know what the probability distribution of the index is. Provided we know the probability distribution of the index, then given the fund value contingent on the index, it is straightforward to obtain the distribution of the fund value. For example, suppose we want to find what value we are going to exceed, say 20% of the time. We would work out the index value which will be exceeded 20% of the time, and then find the fund value corresponding to it. This gives the fund value that will be exceeded 20% of the time. (This is obviously true, provided the higher fund values correspond to higher index values. We shall see shortly this is usually necessary for efficiency.) So it is easy to map from contingent fund values to the distribution for the fund value.

Equally we can go back the other way. If you draw a density function that you like for the fund value, we can just work backwards the other way. Again, we would take some probability, say 20%, locate the fund value we are going to exceed 20% of the time and also the index value we are going to exceed 20% of the time, and then plot that pair of points. Repeating this for all the different probabilities, it builds up the curve of fund values as a function of index ones. So we can always think either in terms of the distribution for the fund value or how the fund value is contingent on the index, and we can map either way. We shall see shortly that this type of framework, at least under simplifying assumptions, is also a condition for the asset allocation rule to be an efficient rule.

3. Efficient diversification through time

This section develops the concept of efficient diversification through time. Some kinds of asset allocation rules essentially throw away money. If you do not want to throw away money, it is important to understand which kinds of rules are efficient and which are not. To get some intuition of how it is possible for a poor dynamic asset allocation rule to throw away money, we will consider a simple mean–variance example that illustrates what can go wrong. The exact relations that need to be followed will be formalized later.

The idea of diversification across securities is now very similar. Here is a very simple numerical example which we will then extend to illustrate time diversification. Suppose we have two securities. Both of them offer an expected risk premium of 10%. Both of them have a standard deviation of 20%, and we will think of this as over a one year period. For simplicity, these securities have returns which are independent. Consider two portfolios, A and B. For Portfolio A we put £100 in security 1 and any remaining wealth is invested in cash. The risk premium we earn is then £10 and the standard deviation is £20. No-one would be silly enough to do that because we know that it is more efficient to diversify. As the two securities are identical the best thing to do is to invest equal amounts in them. A better strategy would be to invest the same amount, £70 say, in each of these two securities, and, again the rest in cash. We shall call this Portfolio B. The variance is now twice 14 squared (from the two securities) which is 392. The standard deviation is £19.8, which is slightly smaller than before, but we have got an extra return premium of £4. We are now earning a £14 risk premium instead of £10 before.

The reason for developing this example in such detail is because we can use exactly the same numbers to tell a different story. Consider an investment in the equity market over two years. In each year the expected risk premium is 10% and the standard deviation is 20%. We know that successive returns are essentially independent so this assumption now looks quite natural. Now see what happens if we have a strategy where we plunge £100 into equity for the first year and then put everything into cash for the second. This corresponds exactly to our Portfolio A: we have got an expected risk premium of £10, and a standard deviation of £20. It would be better to have £70 in equities each year. That way we would still have the same standard deviation for the return at the end of two years but we would have expected to have earned an extra £4 more. Strategies where the manager plunges in and out doing wonderful market timing transactions but actually has no forecasting ability are exactly equivalent to the first very wasteful strategy. You may not realize it is very expensive but we have just shown that it is. Under a market timing strategy of someone who has no forecasting ability, the plunge in and out of equities loses the sort of diversification benefits you would enjoy from a smoother policy.

The magnitudes are significant. If you randomly plunge half the time into cash and half the time into equities, you are giving up 40% of the risk premium that you are earning. You are only getting a £10 risk premium instead of a £14 one. That is an enormous loss.

That is the intuition about why some dynamic allocation strategies are efficient and some throw money away. However, we cannot generally use a mean–variance framework because the pay-offs are too complicated. The mean–variance framework simply helps to develop the intuition. A condition which is often necessary for a dynamic strategy to be efficient, in the sense of not wasting money (the way we wasted £4 on that first strategy) is that the values of the strategy must be path independent. This result, which is not very well known, was first presented by Cox & Leland (1982) in a rather obscure working paper. They show quite rigorously that if we are in a market that satisfies the assumptions for Black–Scholes option pricing (i.e. a constant risk-free rate and constant volatility of the equity market) then a dynamic portfolio strategy must be path independent to be efficient. By efficient we are not referring to mean–variance efficiency, but rather that whatever distribution of outcomes is obtained is purchased as cheaply as possible. (Financial economists refer to this as first order stochastic dominance.) In other words that no money is thrown away.

We describe next what is meant by path independence. Suppose starting in 1992 the index was around 2600. Assume that in 1994 the index is up at 3000 but it got there by one of two paths, either going up to 3600 first or a lower path dropping first to 1900. A path independent strategy is one where the portfolio value in two years time will end up at more or less the same value whichever route the market took to get there. This is an extremely useful result because it means that if we have a path independent strategy then we can characterize it in terms of the contingent profile we described earlier. The very complicated multiperiod problem of what market exposure we should have every day collapses down to a contingent pay-off at a single horizon date. We can rely on option pricing theory to tell us what to do in between.

Thus, the rule for efficiency is that our portfolio value at any future date of interest should just depend on where the index gets to and not on how it got there. If you try and time the market without being able to forecast, that will give a dispersion of fund values at any level of the market. We have seen already that this can be very inefficient. In our example we are losing 4% over the two years, or 200 basis points a year.

Other well-known strategies also suffer from the same problem that they are not path independent and therefore they are ways of throwing money away. A simple one is the stop loss rule. Suppose, for example, I start by investing £100 in the equity market and I am going to leave it there unless the index hits 2000, in which case I will move into cash. If the market reaches 3000 without first falling to 2000 we will have gained 15% and end up with £115. On the other hand, if the market falls below 2000 first, we go down to £77 and then lock into cash, so that at the end we still have £77 plus interest. There is a big divergence between the two ending values even though the market index got back to the same level via a different route. So stop loss strategies are intrinsically wasteful. Dybvig (1988) analyses these various strategies and shows that with a stop loss strategy you could easily lose 80 to 90 basis points a year. That may not seem much money but the risk premium you are getting in the first place is less than 8% per annum, so you are throwing away more than 10% of what you are earning.

A lock-in strategy, where we start off in equities and then move into cash if we hit a level we like, is also path dependent and has a similar level of inefficiency. Finally, another example of a dynamic strategy which Dybvig analyses is to look at repeated short term portfolio insurance. What he does is to look at a strategy of rolling over one year portfolio insurance at the end of each year over a five year period. A lot of funds, particularly in the U.S., have done that sort of thing and that is losing probably about 50 basis points a year.

Dybvig's calculations are not difficult. First you simulate the strategy that a fund manager is using to calculate the probability distribution of future fund values. You then translate this into the corresponding values contingent on the market index. Finally, you use option pricing theory to find out the cheapest way to buy that distribution and then see how much cheaper that is than the amount of money you started with and that led you to the distribution in the first place. The cash saved is a tangible measure of the inefficiency of the original strategy.

To summarize, the rule for efficiency (at least under our assumptions) is that the future fund value should be a non-decreasing function of the future index value. Intuitively it is reasonable that the contingent pay-offs go up because it is cheaper to get money when the index is doing well. Contingent claims that pay off in high index states of the world are cheaper than contingent claims that pay off in low index value states of the world and which provide insurance for those states. So you get as much money as you can when money is cheap and you do not buy so much when it is dear.

I have described a world in which the market index is the only uncertain state variable. This whole theory does generalize into richer assumptions where there are many different state variables, but it gets a bit more messy. We will continue to treat only this slightly simpler framework. We will examine next some other aspects of the robustness of this concept. Clearly it is not entirely robust if we have transactions costs in changing the allocations, though work has been done on this issue (see Hodges & Neuberger 1989). We shall now ask whether the concept is robust if we are thinking of tracking error and whether it is robust if the market mean reverts.

4. Benchmark objectives

If we are interested in tracking error and we set our objectives relative to a benchmark portfolio, then similar results still apply. It is still true that when the index is at a high level, money is cheap and when the index is at a low level, contingent pay-offs are more expensive. We therefore need the profile of our surplus or deficit relative to the benchmark to itself be increasing in the benchmark. The path independence result is still true, so the stop-loss strategy is still inefficient but the way you would modify stop-loss will be different from what you would do under normal risk return criteria.

Benchmarks are hard to justify, except possibly when they represent genuine liabilities to be met. Hodges (1976) compared two criteria: mean–variance tracking error with mean–variance return efficiency using a simple portfolio selection model. As you would expect, each criterion is demonstrably inefficient viewed from the perspective of the other. So if you are pursuing a policy that is mean–variance efficient in tracking error, you are giving up a lot in terms of conventional mean–variance efficiency. This suggests a conflict of interest between the fund manager and the ultimate beneficiary.

Roll (1992) gives a much more formal analysis of optimal tracking error betas. Among other things he shows that optimal tracking error portfolios have betas greater than one, whereas conventional mean–variance efficient portfolios have betas less than one. He suggested constraining beta to a value less than one to enhance mean–variance performance.

This may increase the efficiency somewhat but the result remains sub-optimal. So the bottom line is that tracking error criteria are sub-optimal unless you have a very good reason for saying that that really is what the objective ought to be. If there is a clear liability and you are using the liability as a benchmark, we can say the surplus after we have paid off our liabilities may be a reasonable number on which to use a mean–variance criterion. If you are doing index arbitrage then clearly you are concerned with tracking error. However, it is not a good idea to worry about tracking error just because that is how someone else tots up the points at the end of the year to arrive at a bonus!

5. Market regularities and equilibrium

Most investors seem to like positive skewness. Strategies with pay-offs that are convex from below produce distributions with positive skewness. This is fairly easy to see because a straight line contingent pay-off would essentially be cash plus a static holding in the index and would give a log normal distribution (plus a shift to take account for the cash part). As soon as you add convexity to the contingent pay-off you get a longer right tail and more positive skewness. Conversely, for contingent pay-offs with a sufficient concavity we will get a long left-hand tail and negative skewness. Note also that the slope of the contingent value gives some idea of what the fund exposure is. A portfolio insurance type (convex) strategy gives less exposure at low market levels, whereas a contrarian (concave) strategy gives more exposure at low market levels.

Now the market has to clear, and a paper by Leland (1980) considers who should buy and who should sell portfolio insurance in equilibrium. Suppose we can agree that, other things being equal, most investors prefer positive skewness. In equilibrium we would therefore expect that although people are happy to buy stocks when the market is high, they will need more encouragement to hold them when the market is low. Thus, we would expect higher risk premia at historically low market levels, and low risk premia at historically high ones. Empirical work tends to confirm this view. For example, the variance-ratio tests reported by Lo & McKinley (1988) and Poterba & Summers (1988) suggest mean reversion. Fama & French (1988) have also looked at dividend yields as a way of predicting expected returns and again confirm the same kind of effect. Not all the work is terribly significant statistically. There is a suggestion of possible mean reversion, at or about a five year period, but we do not have long enough data sets to get very good statistical significance over this kind of horizon. However, there is now a large number of studies, most of which tend to confirm this view of the world.

It is interesting how little attention has been paid until recently to the way the market risk premium evolves through time. Even the better finance texts seem not to question whether the market risk premium is likely to be constant through time or how it might change. This has now become a topic of theoretical study. He & Leland (1991) and Hodges & Carverhill (1993) have done work which characterizes the evolution of the equilibrium risk premium through time.

This characterization (which is based on a single representative investor assumption) implies that under quite general utility functions, the longer the horizon the less the risk premium will respond to changes in market value. Figure 1 illustrates this by showing some numerical results that we have calculated using this evolution. The solid line shows a hypothesized relationship between the price of risk, defined as the Sharpe ratio measure of the annual risk premium divided by the standard deviation. The risk premium might be 9% and the standard deviation might be 15% to give a figure of 0.6 for this ratio. We are postulating that it depends on the level of the market index, with a higher risk premium when the index is low and a lower risk premium when the index is high. We start off by assuming that when it is close to the horizon it is going to take the shape of the solid line, and our theory tells us how that will evolve when the horizon is a longer way off. What happens is it flattens out. Whereas it is fairly steep at the horizon, it is getting rather flatter when we are six years away. What that seems to imply is that in a real market, the investors with very long horizons will not require such a big risk premium after the

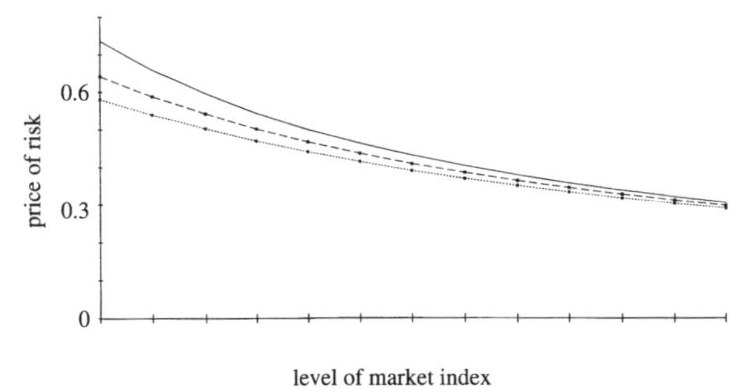

Figure 1. Evolution of the risk premium solution.

market has fallen as compared with the short horizon ones. The market will clear with the investors with long horizons tending to be contrarian, whereas investors with short horizons may feel a need to portfolio insure. We can never say that either a contrarian policy or one of portfolio insurance is the only sensible thing to do. It depends on what your situation is and what your objectives are. We cannot even say that all longer horizon investors should be contrarian, or all short horizon ones should portfolio insure. This is just the balance between the two on average.

The interesting thing about this analysis is that if we are in this kind of equilibrium, where the relationship of the risk premia to the level of the market changes systematically through time, our previous results concerning path independence still hold. On the other hand, there are also more complex equilibrium settings where life is a bit more complicated. We may still be able to use the options framework to think about dynamic asset allocation but the strict form of path independence we had earlier will no longer quite apply.

Tactical allocation

Fama (1991) provides the most recent academic survey on market anomalies, confirming the usual things we all know about: price earnings ratio effects; dividend yield effects; January effects; small firm effects. We know there are studies of in-house analysts that reveal some degree of forecasting skills, and tactical asset allocation policies would aim to be capitalizing on all of these types of things.

Now if we can make forecasts, then again we are making a major change to the Black–Scholes assumptions that Cox & Leland and Dybvig used in their analyses. We can always think in terms of contingent pay-offs as a function of the index. We can also think in terms of option theory to tell us what the exposure should be. However, the Black–Scholes delta is now telling you about our risk-return trade-off, rather than directly about market exposure. Thus, if you forecast, say, that the expected return on the market is 2% higher than normal, you then want to increase exposure by a corresponding amount compared to the calculated delta. This will destroy path independence, but you still have a framework for taking a consistent risk return stance in the market place.

The role of forecasting is particularly interesting. The studies by Hodges & Brealey (1973) and Treynor & Black (1973) look at the relationship between fund performance and the forecasting ability as measured in terms of the correlation between forecasts and outcomes. What they show is that very significant returns can be obtained with remarkably low levels of forecasting ability as measured by the correlation coefficient (R^2). It turns out that if you can make forecasts with an R^2 of 0.01 or 0.02, and use them properly, you can still make quite respectable returns of 2 or 3%. This is a major reason why performance measurement is difficult, because what is significant economically may well not be significant statistically. If you run a regression and find an R^2 of 0.03 you are liable to throw the thing away and say it is no good. That is all you should probably be expecting if you are looking at how good your analysts are anyway. If they are better than that, they will be making a fortune on their own account.

6. Performance measurement

Finally we turn to some issues concerning performance measurement. Most U.K. current practice is rather unsophisticated and the problems are very difficult. The National Association of Pension Funds Committee of Enquiry Report into Investment Performance Measurement (1990) considered, but decided not to pursue, risk adjusted measures. The LIFFE/LTOM (1992) document suggests a sensible treatment for futures but really does not address the issues for options. It treats options as equity substitutes and puts in some sensitivity analysis, but does not really tackle the problems raised by options or by dynamic asset allocation strategies.

Bookstaber & Clarke (1984) have provided analysis which shows clearly that 'methods which depend on mean and variance measures cannot be applied because options strategies mould the return distributions bringing the higher moments into play'. For the distribution of return on an all equity portfolio, there is no problem in using mean–variance analysis. However, suppose the fund manager has written a lot of out-of-the-money covered call options, it pushes the whole distribution to the right and then truncates the right tail. We are left with a distribution which also has a big spike where the call options start to be exercised against the manager, and just by looking at it, you can see that a mean–variance framework cannot be used to compare performance. That epitomizes the problem of performance measurement and it does not matter of course whether the firm actually wrote covered calls or whether it pursued a dynamic strategy which had the same effect. Either way, the probability distribution is quite distorted and we cannot use mean–variance analysis.

We therefore have to relinquish all of the Sharpe (1966), Treynor (1966) and Jensen (1968) philosophy of performance measurement. However, we can still use some of the philosophy of Fama's (1972) decomposition of returns. Fama suggests using benchmark portfolios to attribute performance to various sources. We can have a benchmark that has a moving beta equal to the actual beta of the fund, and the difference between that benchmark and what the fund does you attribute to stock selection. You then compare the moving beta benchmark against something with a fixed beta at the average level, and you attribute that to market timing, and so on. The philosophy can be extended in a variety of ways. One extension is described by Sharpe (1992). He suggests that we calculate the returns on benchmark indices for various asset classes and then run regressions between the fund return and the returns on these indices. That enables us to identify the effective allocation across classes. We can do this in a moving window, so we can track how the allocation across classes is changed. In his paper he contrasts the Trustees Commingled Fund which had a very static mix of asset classes, mostly in small stocks, with Fidelity Magellan

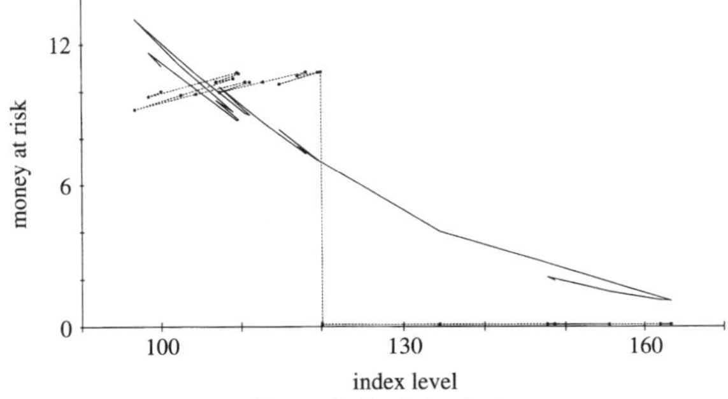

Figure 2. Lock-in strategy.

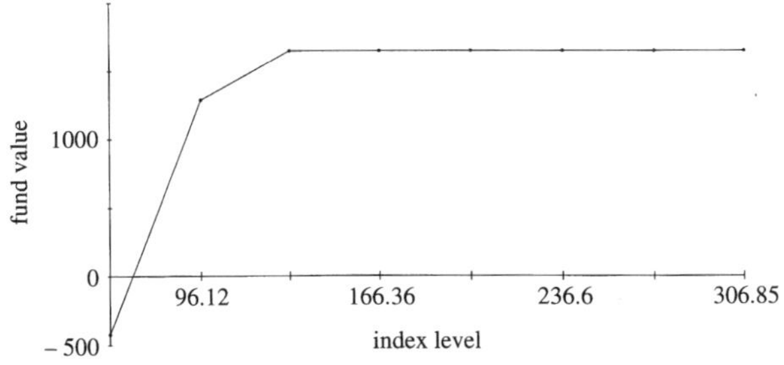

Figure 3. Estimated objective lock-in strategy.

which has an increasing proportion through time in growth stocks and a decreasing one in small stocks.

Finally, Hodges (1991) suggests that we can regress the portfolio exposure on the deltas of options spreads in order to understand horizon objectives and also to understand how close the fund comes to path independence. We create bull option spreads at various market levels and then we explain how close the observed market exposure is to a linear combination of the deltas from each of these bull option spreads. If the residual is zero then the manager is pursuing a classic path independent strategy. If there is a high residual, it is not at all path independent. The analysis gives us directly the inferred future fund value as a function of the index value (i.e. its contingent pay-off). Surprisingly (at least running that with simulated data) this procedure seems to work fairly well. Figure 2 shows a path independent strategy (solid) fitted to an actual simulated lock-in strategy (dotted). They are quite a long way apart, which shows that the lock-in is fairly inefficient. The inferred objectives are shown in figure 3 and you can see the lock-in aspect is revealed from the analysis.

7. Conclusions

In conclusion, the real power of derivatives for fund management lies more than anything for their use in modifying return distributions. We must, of course, recognize the liquidity limitations of the markets, and this conclusion may not apply so accurately to the very largest funds. Inconsistent risk exposure can be very expensive. Market regularities matter, so portfolio insurance probably has a hidden cost of lower expected returns. If you are contrarian you have to be able to bear the risk when the market is bad, but there is probably a reward. For performance measurement, mean–variance measures are inadequate and we are now just beginning to see new approaches that work directly with the pattern of risk exposure through time.

I acknowledge many helpful comments to earlier versions of this paper from Dr Les Clewlow, Dr Michael Selby and other colleagues in FORC, and from participants at seminars in London, Mannheim and Rotterdam.

References

Admati, A. R. & Ross, S. A. 1985 Measuring performance in a rational expectations equilibrium model. *J. Business* **58**, 1–26.

Allen, G. C. 1991 Performance attribution for global equity portfolios. *J. Portfolio Management* Fall, 59–65.

Black, F. & Scholes, M. 1973 The pricing of options and corporate liabilities. *J. Political Econ.* **81**, 637–659.

Bookstaber, R. 1990 *Option pricing and investment strategies* (3rd edn). Probus Publishing.

Bookstaber, R. & Clarke, R. 1984 Options portfolio strategies: measurement and evaluation. *J. Business* **57**, 469–492.

Cox, J. C. & Leland, H. E. 1982 On dynamic investment strategies. *Proc. seminar analysis of security prices*, Centre for Research in Security Prices, University of Chicago.

Dybvig, P. H. 1988 Inefficient dynamic portfolio strategies, or how to throw away a million dollars. *Rev. Financial Stud.* **1**, 67–88.

Fama, E. F. 1972 Components of investment performance. *J. Finance* **27**, 551–567.

Fama, E. F. 1991 Efficient capital markets: II. *J. Finance* **46**, 1575–1617.

Fama, E. F. & French, K. R. 1988 Divident yields and expected stock returns. *J. Financial Econ.* **22**, 3–25.

Galai, D. & Geske, R. 1984 Option performance measurement. *J. Portfolio Management.*

He, H. & Leland, H. 1993 On equilibrium asset price processes. *Rev. Financial Stud.* **6**, 593–617.

Hodges, S. D. 1976 Problems in the application of portfolio selection models. *Omega* **4**, 699–709.

Hodges, S. D. 1991 Ex-post evaluation of dynamic portfolio strategies (or how to tell whether a million dollars has been thrown away). Financial Options Research Centre, University of Warwick (Pre-print 91/23).

Hodges, S. D. & Brealey, R. A. 1973 Portfolio selection in a dynamic uncertain world. *Financial Analysts J.* **29**, 50–65.

Hodges, S. D. & Carverhill, A. P. 1993 The characterization of economic equilibria which support Black–Scholes option pricing. *Economic J.* **103**, 395–405.

Hodges, S. D. & Neuberger, A. 1989 Optimal replication of contingent claims under transactions costs. *Rev. Futures Markets* **8**, 223–242.

Jensen, M. C. 1968 The performance of mutual funds in the period 1945–1964. *J. Finance* **23**, 389–416.

Kritzman, M. P. 1990 *Asset allocation for institutional portfolios.* Homewood, Illinois: R. D. Irwin.

Kritzman, M. P. 1992 Asset allocation for individual investors. *Financial Analysts J.* Jan.–Feb., 12–13.

Leland, H. E. 1980 Who should buy a portfolio insurance? *J. Finance* **35**, 581–596.

LIFFE/LTOM 1992 The reporting and performance measurement of financial futures and options in investment portfolios. *LIFFE/LTOM Recommendations*, January.

Lo, A. W. & McKinley, A. C. 1988 Stock market prices do not follow random walks: evidence from a simple specification test. *Rev. Financial Stud.* **1**, 41–66.

Poterba, J. & Summers, L. 1988 Mean reversion in stock prices: evidence and implications. *J. Financial Econ.* **22**, 27–59.

Roll, R. 1992 A mean–variance analysis of tracking error. *J. Portfolio Management* Summer, 13–23.

Samuelson, P. A. 1969 Lifeline portfolio selection by dynamic stochastic programming. *Rev. Econ. Stat.* **51**, 239–246.

Sharpe, W. F. 1966 Mutual fund performance. *J. Business* **39**, 119–138.

Sharpe, W. F. 1992 Asset allocation: management style and performance measurement. *J. Portfolio Management* Winter, 7–19.

Treynor, J. L. 1966 How to rate management investment funds. *Harvard Business Rev.* **43**, 63–75.

Treynor, J. L. & Black, F. 1973 How to use security analysis to improve portfolio selection. *J. Business* **46**, 66–86.

Trippi, R. R. & Harrif, R. B. 1991 Dynamic asset allocation rules: survey and synthesis. *J. Portfolio Management* Summer, 19–26.

Discussion

R. S. Clarkson (*Dalserf, U.K.*). Experience suggests that an academic analysis of risk is futile in most of the situations. We need to concentrate instead on investor psychology; what Keynes described as 'anticipating the anticipations of others'. As opposed to the second order differential equations we have been discussing, I regard the behaviour of 'good' investors as being at least fourth order in nature, but we simply do not yet have such high order mathematical models. My guess is that in addition to the aggregate uncertainty introduced at the portfolio level by Professor Markowitz more than 40 years ago we need to build in uncertainty at the microscopic level along the lines of the Heisenberg Uncertainty Principle in quantum theory if we are to explain how individual investors behave and thereafter build up a new theory of finance on sounder foundations than those provided by the linear methodologies of the current theory.

There is a story that emphasizes the two crucial components of successful investment: obtaining important information before your competitors and understanding investor psychology. When the Battle of Waterloo was fought in 1815, there were no Reuters or Topic screens to give a real-time account of the engagement. Rothschild, however, had his own private sources of intelligence, and when his messenger brought him the crucial news at about 11 a.m. he went into the gilts market and sold heavily. Others assumed that we had lost at Waterloo, and the market plunged in waves of panic selling. Rothschild bought back at the bottom of the panic and then went out to lunch. When the Government messenger arrived in the afternoon it was realized that we had indeed won the Battle of Waterloo, and prices soared spectacularly. The enormous profits that George Soros made out of currency speculation around the time of sterling's ignominious exit from the Exchange Rate Mechanism suggests that investor psychology has not changed much over the past two or three centuries.

M. A. H. Dempster (*University of Essex, U.K.*). A class of models for dynamic management of portfolios of both assets and liabilities which are capable of capitalizing on even the low levels of accuracy in forecasting returns on securities mentioned – including those for options and futures – are discrete time dynamic stochastic programming models. This type of model was first applied by Bradley & Crane (1972) and Lane & Hutchinson (1980) and modern versions (Dempster & Ireland 1988; Carino *et al.* 1993)) tend to maximize the expected utility of terminal wealth at a finite horizon, taking into account horizon end effects and possibly subject to down-side risk control through explicit (almost sure) constraints. Rate and yield forecasts may be incorporated into the generation of explicit realizations of future rate and yield processes, termed *scenarios*, whose random variations are considered in the optimal evolution of portfolio decisions. More details may be found in Dempster (1994).

Additional references

Bradley, S. P. & Crane, D. B. 1972 A dynamic model for bond portfolio management. *Management Sci.* **19**, 139–151.

Carino, D. R. *et al.* 1993 The Russell–Yasuda Kasai financial planning model. *Management Sci.* (submitted).

Dempster, M. A. H. 1994 Dynamic asset/liability portfolio management. *Mathematical Finance.* (In the press.)

Dempster, M. A. H. & Ireland, A. M. 1988 A financial expert decision support system. In *Mathematical models for decision support* (ed. G. Mitra), pp. 415–440. Berlin: Springer.

Lane, M. & Hutchinson, P. 1980 A model for managing a certificate of deposit portfolio under uncertainty. In *Stochastic programming* (ed. M. A. H. Dempster), pp. 473–495. London: Academic Press.

Index